TURNING STONES

Discovering the Life of Water

DECLAN McCABE

ILLUSTRATED BY ADELAIDE MURPHY TYROL

CAMDEN, MAINE

Down East Books

An imprint of Globe Pequot, the trade division of The Rowman & Littlefield
Publishing Group, Inc.
4501 Forbes Blvd., Ste. 200
Lanham, MD 20706
www.rowman.com

Distributed by NATIONAL BOOK NETWORK

Copyright © 2024 by Declan McCabe
Illustrations © Adelaide Murphy Tyrol

All rights reserved. No part of this book may be reproduced in any form or
by any electronic or mechanical means, including information storage and
retrieval systems, without written permission from the publisher, except by
a reviewer who may quote passages in a review.

British Library Cataloguing in Publication Information available

Library of Congress Cataloging-in-Publication Data

ISBN 978-1-68475-183-9 (paperback)
ISBN 978-1-68475-184-6 (e-book)

∞™ The paper used in this publication meets the minimum requirements of
American National Standard for Information Sciences—Permanence of Paper
for Printed Library Materials, ANSI/NISO Z39.48-1992.

CONTENTS

ACKNOWLEDGMENTS

First and foremost, I could never have completed this project without the patient support of my amazingly smart and talented wife and partner in life, Margaret Vizzard. Our three wonderful children—Heather, Ethan, and Lauren—have also been quite tolerant of my somewhat hermit-like habit of disappearing into the loft of my garden shed to write, edit, read, and rewrite. I'm grateful also for the inspiration that my family has provided for many of the essays.

My colleagues and students at Saint Michael's College have good-naturedly indulged my passion bordering on obsession for all things aquatic. The college's support and encouragement of hands-on science education has fostered field experiences for my students and I that provided much of the inspiration for the essays. I am particularly grateful to my Biology Department colleagues Lyndsay Avery, Donna Bozzone, Paul Constantino, Doug Facey, Ruth Fabian-Fine, Doug Green, Peter Hope, Scott Lewins, Dagan Loisel, Mark Lubkowitz, Carolyn Marsden, Denise Martin, Nicole Podnecky, Brian Swisher, Karen Talentino, and Adam Weaver; kindred spirits all!

Student research collaborators have been central to my work since the 1990s. They have installed equipment, taken samples, and identified thousands of macroinvertebrates. Their work has resulted in professional presentations, publications, a smartphone app. and much of the content of this book. I am particularly grateful to some standout students whose work on aquatic systems has been particularly inspired and inspiring: Alexandra Canepa, Kathleen Coons, Brian Cunningham, Erin Hayes-Pontius, Janel Roberge, and Matthew Toomey.

I have collaborated with educators and researchers too numerous to mention through Vermont EPSCoR (Established Program to Stimulate Competitive Research) since 2008. This group includes professors and administrators from six Vermont institutes of higher education and dozens of high schools in Vermont, New York, Massachusetts, South

Carolina, and Puerto Rico. In particular, I thank Sharon Boardman, Arne Bomblies, Leslie Kanat, Bob Genter, Kathy-Jo Jankowski, Miranda Lescaze, Patricia Manley, Tom Manley, Yiria Muniz-Costas, Janel Roberge, Rosaliz Rodriguez, Sallie Sheldon, Veronica Sosa-Gonzalez, Judith VanHouten, and Lindsay Wieland. This dynamic group of professionals has molded my work, shifted my direction, and introduced me to new fields of study.

I thank Dr. Christina Chant, a biophysical chemist at Saint Michael's College for fact-checking and for making valuable suggestions on the essay "Ice Capades." Thank you, Nathan Buckley, for your wonderful question that stimulated me to write "Submerged Silk Spinners"; keep asking interesting questions.

This book would be incomplete without Adelaide Murphy Tyrol's amazing art. As these essays come out in the newspapers, I am always excited to see Adelaide's unique and creative interpretation of the subject matter. She is meticulous in her attention to the important details, and I was thrilled when she agreed to provide the art for this book.

Two of the essays appeared in *Connecticut Woodlands* magazine and were improved by Timothy Brown's input and edits.

Finally, sincerest gratitude is due to Elise Tillinghast and her amazing staff at *Northern Woodlands* magazine. In 2015, Elise invited me to write for The Outside Story, where most of the essays in this book have been published. I have learned a great deal from the generous and thoughtful edits provided by Elise, Dave Mance III, Cheryl Daigle, and Meghan McCarthy McPhaul. I am extremely grateful for the platforms that the Center for Northern Woodlands Education provides for sharing the natural history and fascinating lives of organisms of all sorts.

This book is part of an education/outreach program funded by Vermont EPSCoR's Grant NSF EPS Award #1556770 from the National Science Foundation. Essays in The Outside Story were sponsored by the Wellborn Ecology Fund of the New Hampshire Charitable Foundation. Additional essays from the Invertebrate Bestiary, a column I write for *Northern Woodlands* magazine, are sponsored by the Lintilhac Foundation. Writing time was generously provided by a Saint Michael's College sabbatical and by National Science Foundation Grant DUE-1742241.

INTRODUCTION

For focus, exercise, and pleasant distraction, I walk the mile-long trail from my home to the Winooski Riverbank in South Burlington, Vermont. By the time I pass a dozen houses, traverse a power-line right-of-way, and take the deep dive through woods to the water, my mind is cleared and worries diminished.

These brief trips provide solitude, grounding, and an opportunity to explore. Slowing down and observing carefully reveals diverse life in this place barely a stone's throw from homes, roads, and an international airport. Each patch of soil, each fallen tree, and every puddle of standing water is a microcosm of life to be appreciated.

During a recent trip, I paused by a river pool protected from the main current by a collapsed piece of riverbank. The river had reclaimed the previously deposited soil and the new pool's surface was already occupied by dancing whirligig beetles. A second group of insects, some nonbiting midges, was bouncing on the water surface, to what end I can only guess.

The bouncing behavior, it turned out, was reckless, because the waiting whirligigs reoriented to each ripple, sprinting, in generally vain attempts to snatch a winged meal. I perched on a passably dry tree root above the pool, watching the unfolding action. After more than fifteen minutes, I was rewarded, or rather a whirligig was rewarded with a midge-sized meal.

I suppose one could measure pool area, beetle and midge densities, and calculate the probability that a beetle might nab a midge, but the simple observation was enough for me to know that it happens. And clearly it happens frequently enough, or the beetles would have evolved to dine elsewhere. Careful observations of natural history are the foundation upon which ecological science is built. And whether you are a scientist, passionate natural historian, or simply enjoy walks in the woods, there are many observations and discoveries to be made.

Fifteen minutes on a tree root was enough for one discovery, but I still wanted to see more than just a single whirligig, fly wings protruding from its mandibles, while swimming through peers jostling for a morsel. I had previously fed ants to water striders, so I proceeded up the riverbank flipping logs and turning stones to find a few. It has been said that bird-watchers leave only footprints while entomologists leave chaos. I was determined to defeat that stereotype, and I carefully replaced each log and rock to protect the invertebrates and salamanders that call them home.

My ant search was brief and futile. But I did find an isopod, or woodlouse, to serve my curiosity. I returned to my pool, the whirligigs briefly scattered, so I sat until they seemed to adjust to my presence. I attempted to gently drop my captive on the water surface as I had previously done with ants, and immediately learned another important lesson. Terrestrial isopods sink! Ants are easily supported by the surface tension, but the isopod fell through to the pool bed without so much as a pause. If the whirligigs noticed, they gave no indication.

The essays in this book are based on my experiences as an aquatic ecologist and naturalist. My interests in the natural world are broad, but I have had a particular fascination with life in fresh water. This fascination dates to a childhood in the Irish midlands and parents who encouraged my siblings and I to pursue our interests. I followed my interests to ponds and the Shannon River. Some of my fondest and most deeply engrained memories are of hours spent with homemade nets in drainage ditches near our home.

Unless you live in a desert, it's quite likely that there is a freshwater habitat close by. Diverse life exists in rivers, ponds, springs, rain gutters, and yes, even in drainage ditches. You don't need specialized equipment or scuba gear to see what amazing organisms live under a stream rock, or in pile of submerged leaves. A simple net and a clear plastic jar can be a window into the exotic world of pond life. Sitting for fifteen minutes on a riverbank can be just as rewarding.

Like any enthusiast, I am close to my subject, sometimes too close to realize how utterly unfamiliar the underwater world is to many. We are, after all, a terrestrial species. But the annual influx of new Saint Michael's College students reminds me regularly that the excitement of discovery is always close at hand, and that for most people, seeing a caddisfly or picking a crayfish from a riverbed is a brand-new experience and a discovery of unfamiliar and exciting life that exists in our very own neighborhoods.

Although I have organized the essays that make up this book into parts, there is no particular reason to read them in that order. If a title grabs your attention, then that is the next one to read. Each essay stands alone, and they are sufficiently brief to read between the busy moments of life. I write to share experiences and introduce the ensemble that occupies puddles, ponds, and backwaters. It is my hope that familiarity can build appreciation, and the next time you visit a pond or stream, you may be tempted to wade in, dip a net, or simply turn a stone.

PART I

WATER FOR LIFE

Water is the driving force of all nature.
—LEONARDO DA VINCI

As you read this book, the signals transmitting words to your brain and conjuring mental images are dependent on water. Each muscular contraction allowing your eyes to track words on screen or page depends utterly on chemistry happening in water. There are other methods to transmit signals as demonstrated by our inventions from telegraph to the internet, but these methods are not the ways of "life." Everything that we consider to be "living" is water-based.

Each of our inventions is a product of minds bathed constantly in water-based blood carrying glucose in and wastes away. You and I are water-based beings, and fully 60 percent of each of our bodies consists of water.

All of the chemistry of living beings including blue whales, desert iguanas, and microscopic bacteria happens in water. Even viruses on the very boundary of what we define as "life" function only after entering a cell that is filled mostly with water. The marvelous chemistry underpinning all bodily functions in all life-forms we have encountered thus far can occur only because of unique physical and chemical properties of the water molecule.

For these reasons, the first part of *Turning Stones* explores just some of the unusual characteristics of a molecule consisting of an oxygen atom bound to two hydrogen atoms. These traits provide constant winter temperatures for some organisms, cap winter lakes with solid ceilings, moderate our climates, and even allow some organisms to walk on water.

The natural processes that clean and deliver our water at zero cost are what environmental economists call "ecosystem services." If we do our part to protect this precious resource, water will continue to sustain not only human lives, but the lives of our fellow travelers on this one finite Earth.

WATER FOR LIFE

Few images have captured the human imagination as thoroughly as the iconic "Blue Marble" photograph taken in 1972 by the Apollo 17 crew while *en route* to the Moon. The view from 18,000 miles up shows the outline of the African continent, the Arabian Peninsula, Madagascar, a cloud-shrouded Antarctica, and the Eurasian landmass wrapping the northern horizon.

But oceans provide the dominant color for the Blue Marble. Water is ubiquitous on Earth with oceans covering fully two-thirds of the planet's surface. The first life-forms and indeed most life on earth evolved in the salty marine environment.

Fresh water is just a drop in the proverbial bucket of Earth's water, but it is essential for our survival, and for the survival of a host of fascinating organisms with which we share the planet. Water containing less than 0.05 percent dissolved salt content is fresh water, more salt than that and the water is considered "brackish." Seawater contains 3.5 percent salt.

Less than 3 percent of Earth's water is fresh water, and two-thirds of that is inaccessibly tied up in ice caps, snow, and ice. After accounting for groundwater and water in the atmosphere, just 0.01 percent of Earth's total water is found on the land surface in lakes, ponds, rivers, and streams. Viewed in the larger context of the Earth and considering accessible fresh water's importance to our economic systems and to our very survival, that surface water is a precious resource worthy of our care and protection.

I am reminded when I wash my dishes that water is considered the universal solvent for good reason. Water on its own easily removes crusted sugary food and dried-on salt. This works because the H_2O molecule has both positive and negative charges. The negatively charged oxygen atom interacts well with some components of the dried-on food and the positively charged hydrogen atoms fare better with other food scraps.

But there's more to the magic of water than just positive and negative charges. Because water molecules are bent at a 104.5° angle with the oxygen atom at the bend, oxygen's negative charge points in one direction. Positively charged hydrogen atoms point in the opposite direction. Each water molecule functions as a very miniature magnet that is attracted to all of its peers, and to other molecules and atoms depending on their charges.

Just as magnets are often labeled with north and south poles, water molecules are considered polar because of the opposite distribution of negative and positive charges. Other polar molecules like sugars and salts dissolve particularly well in water. Nonpolar molecules like oils and greases dissolve poorly in water. Detergent molecules that at one end are attracted to water, and at the other attracted to fatty compounds help solve that problem. Detergents also reduce surface tension, allowing water to better wet clothing in our washing machines.

The polar characteristic of water is important for far more than removing egg yolks and caramelized sugars from our dishes. Water carries all of the raw materials needed for life to and from root hairs deep in the ground to needles as high as 380 feet skyward in the world's tallest tree, a redwood named "Hyperion." Likewise, all the elements and compounds necessary for your life and mine are carried by water.

Water's polarity affects surface tension, heat capacity, and the density anomaly that permits ice to float over constant-temperature water. This book of essays focuses on life in 0.01 percent of the planet's water, and that life, or truly any life, would not be possible without some truly unusual physical and chemical characteristics of the humble water molecule. Essays in this first part of the book cover some of the remarkable consequences of water chemistry that make life on our precious blue planet possible.

SURFACE TENSION

On windless days, a pond's glassy surface is dimpled only by the incessant revolutions of whirligig beetles. Unless of course they are joined by their leggier competitors, the water striders. These distantly related insects together with their predators, the fishing spiders, are perhaps the most charismatic animals living at the air/water interphase. But the total cast of characters, some more aquatic, others more terrestrial, is diverse. In addition to those born to ply the surface, there are those that visit from below to refill their "scuba tanks" or poke the surface with "snorkels," others more tenuously skitter from leaf to stick, and still others appear doomed on contact with the watery surface.

The doomed-from-the-start group include landlubber insects that are blown in by oblivious winds, knocked from a branch, or simply overreaching victims of gravity. Opportunists on and below the surface are ready to avail themselves of these protein-packed snacks. Perhaps only an academic ecologist would view this interface of life and death as a "source of terrestrial supplementation of the aquatic food web."

But how do denser-than-water organisms walk on a pond? The answer is "surface tension," and more particularly, water's unusually strong surface tension. Molecules in liquids are held together by forces of mutual attraction; otherwise, these substances would be gases. And when we heat liquids above their boiling points, the energy we supply breaks these forces of attraction and liquids become gases.

Forces of attraction deep in liquids are shared among molecules in all directions. However, at the surface there are no liquid molecules above, and so all forces of attraction are devoted below and horizontally along the surface. As a result, liquid surface molecules are more strongly attracted to each other than would otherwise be the case and liquid surfaces are skin-like and act like mini trampolines, or major trampolines in the cases of lakes and ponds.

Surface tension strength, and the characteristics of objects, like insect legs, determine whether the surface breaks and the insect gets wet or remains high and dry. Engineers designing pumps to move various liquids use tables of surface tension, measured as dynes per centimeter, or the energy required to break a surface film of 1 cm in length. Ethanol, the alcohol found in beer and wine, has a surface tension of 22 dynes/cm while gasoline measures 19 dynes/cm; water comes in at an astonishing 73 dynes/cm.

To develop a deeper appreciation for what these numbers mean, I ran a few simple experiments. I dripped liquids from an eyedropper onto a polyester jacket hung on my office door. Water bounced right off, and it took repeated drips on the same spot to wet the sleeve. Alcohol instantly sank into the fabric without a drop repelled. I didn't mess with gasoline; an office smelling of alcohol first thing in the morning was enough to explain.

To get to the heart of liquid surfaces supporting dense objects, I decided to use steel. Steel is far denser than any of the liquids I had on hand and seemed an ideal test substance. I punched a few staples from my stapler being careful not to bend them. I weighed one; it was 0.034 grams, about as heavy as ten to twenty mosquitoes or as much as a large water strider.

I placed a staple on a piece of tissue paper and dropped it on the surface of water in a coffee cup. Once the tissue was thoroughly wet, I gently pushed it under leaving the staple on the water surface where it remained for more than an hour. Great, water has strong surface tension, but what about other liquids?

To see if water was really "all that," I compared it with alcohol. I repeated the experiment; the tissue saturated immediately and went below bringing its staple along for the ride. I repeated with folded layers of tissue, which slowed the process, but the staple still sank.

My morning's tabletop experimenting convinced me of the strength of water's surface tension, and countless encounters with water striders, emerging mosquitoes, springtails, and fishing spiders have more than convinced me of the importance of the community ecologists refer to as the "pleuston," or those organisms living on, or in the case of some snails, crawling beneath the surface tension.

To experiment on a liquid with a stronger surface tension than water, I borrowed a bottle of glycerin from a colleague; you can find it at the pharmacy. Its surface tension is just higher than water's at 76 dynes/cm. I placed a staple on the surface using tweezers. The staple stayed right on the surface. I went to lunch; it was still perched on top when I returned and went under only when I added detergent to break the surface tension.

Water, of course, has a great many other characteristics that make it perfect for supporting life on our planet. And glycerin is certainly not something I'd like to encounter in a lake or pond. Water's surface tension just happens to be a characteristic that physically supports life and provides yet another ecological niche where various life-forms found in the second part of the book can ply their trades.

AN EXTRAORDINARY
FONDNESS FOR HEAT

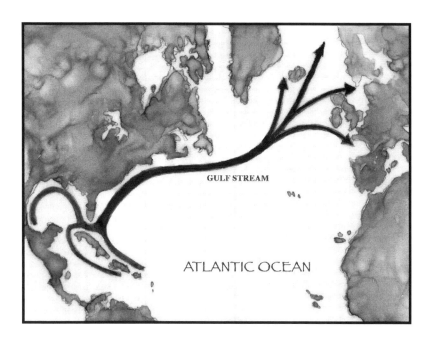

I grew up on the 53rd parallel, as far north as Newfoundland and
Labrador, Edmonton in Alberta, and Barnaul in Siberia. Despite
being six hundred miles closer to the Arctic Circle than my current
Vermont home, Ireland's climate is comparatively balmy. Snow is so rare
that it cripples the country as it might a southern US state ill equipped
for layers of white. By February, when I'm still shoveling snow, my Irish
siblings will post photographs of crocuses and snowdrops, I post mine
in April.

Ireland's mild climate is a result of water's incredible capacity to
store heat. Ocean water warmed by the Gulf of Mexico's subtropical
sun flows northeast along the United States and Atlantic Canada coasts
before feeding the North Atlantic current that bathes Europe's coast.
Warmth carried by these ocean currents moderates Western Europe's

climate and spares Ireland the worst rigors of winter at 53 degrees latitude.

Inland lakes have similar effects. I recorded the maximum and minimum temperatures for towns along the historic Erie Canal. Canal routes are fairly flat, and this one runs east and west across New York State thus minimizing both elevation and latitude impacts on temperature. Starting from the west where Lake Erie moderates the climate, Buffalo's average low is 19 degrees Fahrenheit and the high is 79, giving an annual temperature range of 60 degrees. Traveling east, Rochester, moderated by Lake Ontario, has an annual range of 65 degrees. Continuing, and departing now from lake influence, Syracuse has a 67-degree annual range. At 55 miles from Lake Ontario, Utica's range is 71 degrees, and another 23 miles gets us to Little Falls, where the range is 72 degrees.

The pattern reverses as we approach a larger and more influential water body, the Atlantic Ocean. Schenectady has a 70-degree range and Albany, just 110 miles from Long Island Sound has a 69-degree range. Abandoning the Erie Canal and continuing due east to the Atlantic, Boston has an annual range of 60 degrees, just like in Buffalo, but Boston, influenced by the much warmer Gulf Stream, averages 3 degrees warmer.

These climatic influences are due to water's unusually high "heat capacity," the amount of heat needed to increase the temperature of a substance by a certain number of degrees. Heating metals like lead or gold up by a few degrees requires little energy but heating water by the same amount requires more than 30 times the energy. Water resists changes in temperature, and once heated, it efficiently stores heat.

It is because of water's heat capacity and availability that we use it in car cooling systems. It is mixed with antifreeze to protect engines in cold weather and to raise the boiling temperature. Water absorbs the engine's heat and releases it through the car's radiator and the heating system that defrosts your windshield and warms your toes.

High school participants in Vermont EPSCoR's (Established Program to Stimulate Competitive Research) outreach program install temperature sensors in Vermont streams. They zip-tie sensors to metal rods driven into creek beds, where they record water temperature every two hours. Students preparing data for final presentations witness stark differences between water temperatures and air temperatures taken the same day.

When Rutland High School recovered their probe from a Poultney River tributary in 2017, they learned that the river water temperature reached a high of 70 degrees in late September and a low of 43 degrees by the end of October. The air temperature in Rutland at the time peaked at 90 degrees before falling to a low of 28. Because of water's resistance to temperature change, the river ranged over just 27 degrees while the air temperature spanned a full 62 degrees.

So reliably consistent is this difference between air temperatures and water temperatures that scientists can use it to determine if and when ponds or reservoirs dry up. The Vermont Center for Ecostudies installs temperature probes in the deepest part of each vernal pool they monitor. Vernal pools dry out each year and host a fascinating array of organisms. When a probe temperature in a particular pool starts to match air temperature, they know the pool has dried out, and is therefore truly a vernal pool.

Water's high heat capacity has implications beyond moderating terrestrial climates and utility in vernal pool monitoring. Thermal stability in aquatic systems provides a safe haven from weather extremes for all members of aquatic communities. Fish, macroinvertebrates, and aquatic plants are spared the dramatic temperature swings endured by terrestrial organisms. And because cooler water holds more oxygen, some fish and macroinvertebrates survive only in the coolest of streams.

Stream-side trees further stabilize stream water temperatures by shading streambeds and reducing solar heating. Well-designed stormwater-retention ponds that contain and slowly release stormwater draining from roofs and parking lots also play a part in protecting streams from heat extremes.

A Saint Michael's College student proposed measuring water quality from the outflow from South Burlington, Vermont's, stormwater-retention ponds. We visited the sites after three different rainstorms, but the ponds never overflowed. Water instead infiltrated to the groundwater, where contact with soil and rock would reduce its temperature before it reached the nearest stream.

Whether you live north of 53 degrees or farther south, water influences your climate. And you in turn can influence the temperature of waters hosting your favorite fish. Thoughtful development and some judiciously placed trees can be the difference between a trout stream and just another stream.

ICE CAPADES

Come mid-January, when I'm acclimatized to Vermont's winter, I enjoy an occasional stroll on the icy surface of Lake Champlain. I favor bays sheltered from the brunt of winter winds where the ice has had ample time to thicken. I pull microspikes on over my boots and off I go.

There's room to roam between Burlington and the breakwater that parallels the shoreline. The lake ice locks spectacular natural art in place. Bubbles trapped under December ice are entombed as January's ice forms below. Crystalline patterns resembling minute stars form during the various freezing and thawing cycles that occur as lake ice interacts with fallen snow.

While the winter air dries our skin, aquatic life goes on under the protection of the solid barrier—life that is possible because of a strange quirk in the physical chemistry of water. Most liquids shrink as they cool and eventually, they become solid, and the solid form continues to shrink as it continues to cool. The old mercury thermometers that have fallen from favor took advantage of this principle.

And if you used water to make a thermometer, it would work quite well, until, that is, the water cools to 39.16 degrees Fahrenheit (4 degrees Celsius). Below that, water starts to misbehave, or at least behave differently than most other liquids. At 39.16 degrees, water shrinks to its most compact and dense form; any cooler and it starts to expand again and continues to expand as ice crystals form. This phenomenon is called the "density anomaly of water."

It is because of this strange expansion that ice cubes in your drink float. Because ice floats, our lakes freeze from the top and are frozen only near the surface. Sunlight and rising temperatures can thaw them from the top in spring. If Lake Champlain froze from the bottom to the full extent of its four-hundred-foot depth, it would be impossible for sunlight and warm air to have much impact and most of the lake's water would never thaw.

Lakes and seas that froze from the bottom would also mean that life on the bottom would have evolved very differently, if at all. Fish would have evolved ways to survive freezing or would be extinct. The insects that spend winter months fattening up for a spring or summer hatch would not have that feeding opportunity. It is not an exaggeration to say that some strange chemistry of water has permitted life as we know it.

It is worth considering, therefore, what it is about water that makes this possible. Water, or H_2O, is two hydrogen atoms attached to an oxygen atom by strong covalent bonds. These three atoms form a shape like a stubby boomerang, with the oxygen at the bend and the hydrogen atoms forming two arms. The boomerangs fly around in liquid water connecting and breaking weak hydrogen bonds as though the molecules were in a three-dimensional game of tag. As the water cools, the boomerangs fly closer together and the water literally shrinks.

At just under 40 degrees Fahrenheit, when the water is as dense as it can get, it stops shrinking and starts to expand again as it approaches freezing. At water's freezing point, the hydrogen bonds become less transient and lock the boomerangs into slightly buckled hexagonal rings and, attached together, the rings look like a sheet of chicken wire. The rings have space in the middle, and so instead of being packed close together like in liquid water, each water molecule is close to just four adjacent molecules. The spacing of the molecules is so great that water expands about 9 percent as it freezes, making ice buoyant on water. Umpteen identical sheets of these hexagons are bound together, face-to-face, building thickness until a near infinite number of hydrogen bonds can support ice shanties, trucks, and people out for a ramble.

Something to consider as you bait your hook and drop a line through the ice—life started out in water more than three and a half billion years ago and only came ashore half a billion years back. How different might things be if it weren't for the strange behavior of the humble water molecule?

LIFE AT 39 DEGREES

On a picture-perfect winter morning in 2019, twenty Saint Michael's College students and I visited Vermont Fish and Wildlife scientists for ice fishing at Knight's Point on Lake Champlain. We drilled holes, baited hooks, learned about ice safety, and identified fish—and even caught a few.

The ice we tentatively walked on provides unshakably constant temperatures for those living in the water below. Burlington's February 2019 air temperature ranged from 72 degrees Fahrenheit to negative

30, but water temperatures in Lake Champlain fluctuated a mere seven degrees, from 32 just beneath the ice to 39 degrees at depth. This aspect of the under-ice environment never changes; 39-degree water remains a winter constant unless lakes freeze solid or stop freezing altogether.

Water is most dense at about 39 degrees Fahrenheit. When it cools below that temperature, it expands, which is why ice floats (see "Ice Capades"). In frozen lakes, the coldest water remains just beneath the ice where, on very cold days, that water freezes and adds to the thickness of the ice. Meanwhile, the denser 39-degree water sinks below this near-freezing layer and extends all the way to the lake floor. Fish, insects, amphibians, and a few brave mammals are exquisitely attuned to this winter reality.

The creatures who survive under the ice have evolved over eons, fine-tuning their physiology and behavior to thrive at that specific 39-degree mark, which remains a winter constant from year to year and lake to lake. Just as our enzymes work best at a body temperature of 98.6 degrees, fish enzymes work best at fish body temperatures, which change with the season. As lake water cools, fish stop making enzymes that worked in summer temperatures and start producing winter versions.

Different fish species take different measures to survive this chilly winter water. Some, like yellow perch and largemouth bass, adjust by slowing their activities, metabolisms, and need for food. Others, like northern pike, remain more active. But even less lively fish can still be tempted by a tasty morsel; ice fishing in the North Country may yield yellow perch, pike, salmon, trout, walleye, and rainbow smelt.

Smelt have an unusual ace up their fishy sleeves for survival at low temperatures: antifreeze. As temperatures cool, the smelt produce increasing quantities of glycerol. Combined with antifreeze proteins, glycerol keeps smelt moving, even at temperatures below 32 degrees. This trait is useful for smelt populations that migrate to sea, where the salt water freezes at 28 degrees. It also explains why a baited hook dropped through a hole in lake ice can land smelt in even the coldest conditions. And for those lucky to catch enough smelt for a meal, the glycerol contributes to the sweet taste of this fish.

Many of New England's frogs also overwinter under ice. Contrary to popular belief, few frogs hibernate buried in mud. These amphibians absorb oxygen directly through their skin, an impossible feat in anoxic mud. Frogs are typically found on top of the lake or pond floor, and often near inflowing streams and seeps, where currents deliver oxygenated water. Although frogs in winter ponds cease feeding and slow down to conserve energy, if stimulated they can still move and swim.

Springtime melt brings a gradual transition from ice to open water at a rate that allows organisms to adjust. Once water warms from near freezing to 39 degrees, density differences disappear, resulting in a fleeting, uniformly warm water column. Gradually, the surface water will warm even more, but there'll still be plenty of cold places at depth. Life within the lake adjusts to the changes.

As the water warms, frogs stretch their legs once more, surface for air, and return to full activity. Fish stop making cold-optimized enzymes and switch to summer equivalents, and they increase their foraging.

Some fish, however, continue to seek cooler waters even in the heat of summer. Rainbow smelt eschew warm shallow water and migrate to deeper, cooler haunts. More than 90 percent of the 1,600 smelt that University of Vermont researchers netted during a 2007–2008 study came from 60 feet below the surface or deeper, where the water temperature was 45 degrees and cooler, even in midsummer.

And what of our winter ice fishing adventure? Twenty students fishing for two hours yielded three yellow perch, none of edible size and all dutifully returned from whence they came. It seems we are little threat to Lake Champlain fish stocks and should be grateful for a well-stocked cafeteria.

WATER CLEANED FOR FREE

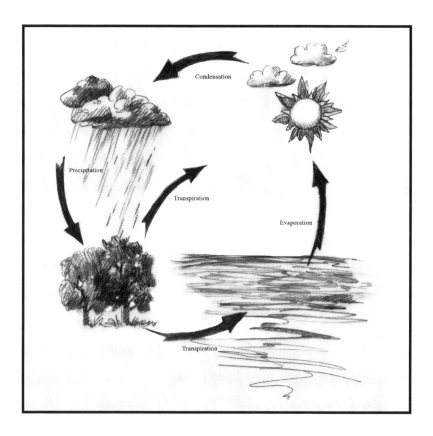

My family and I visited the site of Ben Franklin's home, an obligatory stop when touring Philadelphia. On the pavement by the metal-frame model standing in for the Franklin home, there's a label that says, "Franklin Privy Pit," and another reading "Read Franklin water well." It appears that the Franklin/Read home had all of the modern conveniences of their time.

I was struck by the short distance between these essentials of eighteenth-century life. The privy pit was sited scarcely a dozen feet from the well. One might wonder what a preeminent scientist like Franklin was thinking. However, it's worth noting that the privy is dated in 1787, long before John Snow's work on Cholera in London's wells and ninety years before pasteurization. Some latitude is due.

Before electric water pumps, privies next to wells were commonplace. Both were needed conveniently close to dwellings, and it rarely caused problems. Soil filters most particles from water moving through, including particles as small as bacteria. This is why wells and springs can deliver drinkable water.

Water treatment flips this principle on end, passing water vertically through sand filters. In water treatment, added chemicals attract small particles together making bigger, easier to remove particles and we regularly reverse water flow to backwash the sand and remove accumulated particles.

Soil is not the only way nature cleans our water for free. We learn the water cycle in grade school: evaporation, condensation, precipitation, percolation. Water endlessly cycles among land, sky, and ocean in a solar-powered dance choreographed by climate and the physical properties of water.

Groundwater moving slowly through soil and rock certainly loses solid particles, but it gains dissolved salts, in quantities determined by the rock type. In limestone, water gains a lot of calcium and magnesium; these dissolved minerals reduce lather formation when we use soap, and this water is considered "hard." Water softeners remove these minerals, reduce sedimentation in pipes, washing machines, and dishwashers, and help us lather up.

If we consider all of the rivers flowing into the oceans, a predictable quantity of dissolved salt flows to the sea each year. Evaporation from the oceans leaves this salt behind and essentially salt-free water falls back on our landscapes.

In 1899, it occurred to John Joly of Trinity College Dublin that if we know the annual total of salt entering the ocean from rivers, and if we know the ocean's salt concentration, then we should be able to calculate how long these rivers have been flowing and therefore the minimum age of the Earth. His 80-to-100-million-year estimate more than doubled contemporary scientific estimates and certainly far exceeded Bishop Ussher's biblically derived number famously quoted in the Scopes Monkey Trial. Joly's number, however, fell far short of current estimates of 4.5 billion years.

However, salt, or sodium chloride, the table salt we sprinkle on French fries, is a limiting factor for water and wastewater treatment. With each passage of water through our homes there is an incremental increase in sodium chloride concentration that our sewage plants can't easily remove. Agricultural irrigation and even lawn watering wash additional salt from the landscape into rivers. Salt increases so much in the Colorado River as it progresses downstream that international agreements are needed to ensure that water remains suitable for downstream needs by the time it reaches Mexico.

The water cycle deals with salt using the sun's energy across the entire ocean surface. Water evaporated from this vast expanse condenses and falls from the sky. Planet-scale distillation delivers clean water supplies over land.

Naturally produced clean water depends on our responsible use of resources. If we release mercury into the air from smokestacks, then the "clean" rain will include mercury. If we allow our dogs to defecate on paved surfaces, we can expect bacterial contamination and beach closings downstream, and if gasoline and heating oil leak from aging tanks, then our well water will be undrinkable. One truly remarkable example puts all of these principles together to deliver safe drinking water to more than eight million people.

New York City's famously reliable water supply is the poster child for watershed protection. For more than a century, one of the largest cities on the planet has provided safe drinking water for its population just by chlorinating water piped from reservoirs in the Catskill Mountains. Stakeholders from New York City, the Catskill region, and New York's Department of Environmental Protection have implemented a program of septic tank replacements, land acquisition, wastewater treatment plant upgrades, and whole-farm plans addressing agricultural sources of pollution. All of these efforts combine to successfully protect watershed integrity. Only in 2015 did New York City open its first modern water filtration plant but 90 percent of the city's drinking water is still unfiltered, and the reservoirs connected by enormous tunnels to New York City also provide recreational opportunities and support healthy ecosystems with rich fish communities.

We are not returning to the low population densities of Ben Franklin's day, and we still need clean water. We have advanced far past the state of the art in Franklin's time. The Catskills watershed protection demonstrates how we apply that knowledge. With careful planning and wise use of resources, our waterways can provide swimmable, drinkable water for our use while also supporting natural fish and wildlife communities.

PART II

LIFE ON TOP

You gotta keep trying to find your niche and trying to fit into whatever slot that's left for you or to make one of your own.
—DOLLY PARTON

This quote from Dolly Parton, perhaps aimed at musicians, is applicable to all who strive make their way in the world, human or otherwise. We all need resources to live—food, water, and shelter at the most basic level.

When resources are in short supply, organisms are forced to compete for them. Even when competition is as subtle as one plant intercepting light with higher leaves, there are winners and there are losers. And losing the competition for light, water, or food has dire consequences.

One way that organisms minimize competition is by finding "whatever slot that's left," moving into spaces in the environment with fewer competitors, or as biologists put it, by entering an "unexploited niche."

Because very specific adaptations are needed to live on top of the water surface, few organisms can use this particular niche. For example, of the roughly one hundred beetle families in North America, only the *Gyrinidae* live on the water surface, and even this family only uses the surface as adults.

In this part of the book, we will explore life on the water surface. What is it about this uniquely adapted set of organisms that allows them to flourish where few would dare to tread? How can a water strider live on wide-open water with potential predators seeking prey from above and below? How can denser-than-water bugs walk on water absent divine intervention, and how can ripples signal death for some organisms and communicate "meal time" to others? And how can a male spider communicate his intentions to a far larger female without becoming her next meal? These questions and more will be addressed in this part of the book.

LIFE ON TOP

If you have traveled in a watercraft, you have utilized buoyancy to stay afloat. Waterbirds including ducks, swans, and geese together with water lilies and the duckweed discussed in this part of the book similarly takes advantage of being less dense than water to occupy their particular niche on the water surface.

But other organisms, including some heavier and denser than water, use an entirely different principle to stay above or suspended just beneath the surface. The surface tension discussed in the previous part of the book is a force with little impact on human-sized organisms. But at the scale of an insect, surface tension is used by some specially adapted organisms, and is a force for destruction of others.

Organisms living on the surface of water belong to a community called the "pleuston." Some organisms, including duckweed, use gas-filled spaces to achieve true buoyancy, but many use hydrophobic hairs and other structures to repel water and keep their denser-than-water bodies high and dry while riding the surface tension. Others such as mosquito larvae use water-repellent snorkel structures to suspend themselves beneath the surface while breathing air from above, and still others like diving beetles come to the surface to refresh their air supplies before submerging once more.

The amazing adaptations that make possible surface skating, snorkel breathing, and transportation of bubbles below the surface work in large part because of water's incredibly high surface tension. For some species such as mosquitoes, regularly occurring ripples and waves are enough to disrupt the relationship with surface tension, but others including whirligig beetles and water striders seem to take waves in stride. In fact, more than forty species of water striders called "sea skaters" are among the few insects to colonize marine environments. Five of these sea skaters are even found far from the coast on the open ocean.

There are risks from above and below the water for organisms committed to life on the surface, but clearly the benefits have outweighed the risks, because many of these organisms still exist and have not succumbed to displacement or extinction. Moreover, exquisite adaptations have evolved to increase the fit between these organisms and their razor-thin niche.

This part of the book covers organisms that use the surface as their primary home. I have included mosquitoes in this part because each life stage depends on surface tension for success. It may be helpful to reference the essay on surface tension in the opening part of the book while reading the essays in this part.

FOUR EYES ON YOU

"Wʜat's this shiny black beetle with four eyes?" asked Erin Hayes-Pontius, a visiting University of Vermont student, while seated at her microscope. Without glancing up from my own scope I answered, "that's a whirligig beetle." Erin's answer came back: "err, cute . . . but what's it really called?"

I grant that the name whirligig is a bit odd—particularly when applied to an inert pickled beetle—but there are excellent reasons for it. In life, whirligig beetles weave and whirl on pond and river surfaces among dozens of their peers. They move like miniature motorboats that appear to lack rudder function. There's method to this seeming madness. The mesmerizing movement confuses predators, who find it difficult to focus on any one individual. Ecologists call this phenomenon "predator dilution." It's like the old joke about the two friends and the tiger: "I don't need to outrun the tiger, I just need to outrun you!"

Whirligig beetles have other tricks that reduce their likelihood of becoming fish or bird food. In common with many aquatic insects, they use countershading to blend with their environment; they are black against a dark background when viewed from above, and their underbelly is pale against the sky when viewed from below.

The four eyes that Erin noticed are a unique feature of the whirligig beetle family *Gyrinidae*, at least among insects. All it takes is a quick look with a magnifying glass to notice that these beetles have two fully formed compound eyes looking up at the sky, and a second fully formed pair looking down below the waterline. There's a dark band of exoskeleton along the side of the head that externally separates upper from lower eyes. This must be an incredible defensive asset against aquatic and aerial predators, and as a collector of insect samples, I can certainly confirm that when I swing my net from above, whirligigs dive. How their sensory systems make sense of the dual images pushes the limits of my imagination. Perhaps it's akin to seeing and hearing at the same time?

Artem Blagodatski and his colleagues of the Russian Academy of Sciences used an atomic force microscope for a deeper gaze into whirligig eyes that revealed fascinating adaptations to the beetle's dual environments. The surface of the underwater eye is smooth and optically attuned to the refractive properties of water. But because light travels differently through air than through water, the above-water eye surfaces are roughened by maze-like nanostructures that reduce reflectance and improve performance and predator detection above the water.

Yet another defense mechanism involves the production of a distasteful compound that reduces a predator's interest in dining on whirligig tartar. According to a paper by Bernd Heinrich, Professor Emeritus at the University of Vermont, and Daniel Vogt, Professor Emeritus of Biological Sciences at the State University of New York in Plattsburgh, naive fish may take their chances on a whirligig beetle, but even a single exposure is enough to cause intense food aversion. The beetles are therefore quite safe from fish despite the occurrence of large numbers of beetles swimming completely out in the open on many water bodies. In fact, the large numbers may well serve to advertise to fish that these beetles do not belong on the menu.

While whirligigs are most obvious as adults on the surface, life for them begins underwater. Female whirligigs deposit their eggs below the water surface on the stems of vegetation. The predatory larvae dine on small invertebrates on pond and stream floors. The resemblance between larval whirligig beetles and hellgrammites is uncanny. In perhaps the strangest case of convergent evolution, both of these organisms have two pairs of hooks at the back end, lateral filaments that in whirligigs at least function as gills, and impressive mandibles at the front end. It's no wonder my students often think they have found a slimmed-down hellgrammite when whirligig larvae show up in our samples.

You may wonder why whirligigs spend so much time on the surface, when many aquatic invertebrates including youthful whirligig beetles live on the bottom. What they're doing is exploiting a niche by preying on insects that find themselves stuck in the surface film. Whirligigs use a combination of visual cues and water surface vibrations to locate prey. They then circle their victims before using front legs that are typically tucked into streamlined grooves to grab them.

Whirligigs eat several soft-bodied prey species found on the water surface. Their diet includes flies and springtails, but they also eat mosquito larvae that come to the surface from below to breathe. Emerging adult mosquitoes balanced precariously on their pupal exoskeleton must also be sitting ducks. I can't help but wonder how many mosquito bites I must have avoided because some hungry beetles intercepted the buzzing pests before they ever took flight.

Next time you see a flotilla of rapidly moving, chrome-domed beetles at your favorite fishing or swimming spot, you might perhaps thank them for their pest-control services. And remember that, although they may appear to be ignoring you, they likely are keeping an eye—perhaps four of them—on you.

THE GREAT DUCKWEED MIGRATION

The word "migration" conjures images of vast wildebeest or pronghorn herds crossing plains in unison, or hummingbirds traversing the Gulf of Mexico. When charismatic birds leave our New England forests, migration is typically the explanation. But how can a group of plants disappear, without discarding leaves, stems, or other evidence of their presence?

Duckweeds are in the subfamily *Lemnoideae* and are the world's smallest flowering plant. Their small oval leaves float on ponds and quiet backwaters. Rootlike fibers dangle in the water. Although I'd noticed them on Saint Michael's College experimental ponds, I'd never paid close attention to them—until they disappeared.

Two years ago in October, my Saint Michael's College students and I visited the ponds and observed that they were densely carpeted with floating duckweed, but when we returned in November, they were gone. A few dead leaves did not explain this dramatic loss. In spring of the following year, this magic trick played out in reverse: mid-April, zero duckweed; early May, bank-to-bank coverage. Winged migration seemed unlikely. I was baffled and intrigued.

Last fall, I regularly visited the ponds to get to the "bottom" of the mystery, and this insect guy learned what aquatic botanists already knew. As fall progresses, duckweed leaves gradually thicken and sink below the water's surface. Fallen leaves obscure the plants on the pond floor, where they lie safe from the damaging effects of ice.

During the following year's spring melt, I visited the ponds daily and observed how the duckweed popped up again across the pond surfaces. They emerged shortly after the spring peepers. I imagined that the frogs had sung them to the surface.

What actually causes them to sink, and later to surface, is less romantic than the call of a frog. It all comes down to density, buoyancy, and some tricks of plant physiology. Duckweed leaves float because of air pockets between their cells. As fall progresses, the duckweed in the college ponds, *Lemna minor*, accumulates starch in its leaves, filling up the air pockets and increasing plant density. Eventually, the plants sink. But how do they come back up?

Mid-April is peak season for duckweed reappearance in our ponds, although some stragglers continue returning later. The plants have arrived en masse, just like the swallows to San Juan Capistrano. On April 17, I netted some floating duckweed and some still-sunken duckweed from the pond floor. Nearly every floating plant consisted of three leaves: a larger, darker leaf that tended to hang just below the pond surface, and two smaller, vibrantly bright green, more buoyant leaves growing from its edge.

The still-sunken plants had larger, darker leaves and less developed bright green leaves. I put some of these plants in glass beakers on my office window ledge to watch them develop. Sure enough, within a day, the first plant came to the surface. The growing bright green leaves were serving as the plant's water wings.

I was curious about the starch, and a quick splash of iodine told the story. Iodine turns starch a blueish black. The sunken plants were full of starch. In the floating plants, starch had migrated from the old leaf to the new sprouts, which also had air pockets. It seems that overwintered leaves provide starch to the new spring generation whose metabolism and growth produce enough carbon dioxide to float them to the surface. Photosynthesis-producing oxygen then helps keep them afloat.

I'll admit, this process lacks the drama of wildebeests, but it is a form of migration, measured in feet and inches. My ponds are small, plastic, and quickly warmed by the spring sun. Natural ponds are deeper and warm more slowly, so there may be time to witness the return of duckweed "herds" this May in a pond near you.

MOSQUITOES—
LIFE UNDER TENSION

A good friend was in touch; her son was enduring allergic reactions to mosquitoes, and, like any good parent, she sought solutions. I told her that the most practical, nontoxic way to deal with the problem was to consider a mosquito's life cycle, and to interrupt it where it starts.

Mosquitoes begin their lives as eggs laid singly or in rafts, in most cases on the surface of water. We purchase mosquito egg rafts at Saint Michael's College to run student experiments with hatchling larvae.

A female mosquito, potentially using your blood or mine for energy and protein, delicately alights on the water to lay her eggs. Humans, operating at entirely different scales, fail to alight on water; we break through the surface and, if all goes well, we float. Alighting is not the same as floating—in fact many insects such as water striders are denser than water and therefore cannot float. Rather, insects are held up by surface tension.

Water molecules pull together, as you can witness when water beads up on a waxed surface. In a pond, or droplet on your tent fly, water molecules are more strongly attracted to each other than to gas molecules in air. Forces of attraction in water are strong and provide a skin-like structure on water that can support a small insect.

As you may have read in the opening essay of this part of the book, surface tension is strong enough to hold up steel. To see, lay a sewing needle flat on tissue paper; lower the tissue into water and let it saturate and sink. If all goes as planned, the needle will remain on the surface. And steel needles do not float; prove it by touching it with your finger. When your salty skin breaks the surface tension, the needle will sink like any well-behaved piece of steel.

Mosquitoes are neither dense as steel nor absorbent like tissue paper. Their "feet," or tarsal leg segments, repel water. Professor Wu and colleagues from Dalian University of Technology in China found that mosquito leg scales have nanoscale ridges and cross ribs that allow them to safely land and take off from the water surface. So hydrophobic are these scales that Professor Wu calculated they could bear more than twenty-three times the weight of a mosquito on the water surface, making a water strider's legs, which can support only fifteen times that insect's weight, look positively wimpy.

When mosquito eggs hatch, the larvae, or "wrigglers," are vaguely tadpole shaped but smaller. They breathe air through a rear-mounted snorkel or siphon. The top of the siphon has five flaps that, when submerged, close to form a protective cone that keeps water out. When the siphon breaks the water surface, the flaps open to form a floating triangle shape from which the resting larva hangs. Like other arthropods, growing larvae shed their exoskeletons several times as they grow from hatchling to full-sized larva. And then they transform into pupae that are unique among insects.

When I think "pupa," I remember my children's copy of Eric Carle's *The Very Hungry Caterpillar*. The pupa, or cocoon, sits quietly until a spectacular adult emerges. Mosquito pupae hang from the surface tension, but at the first hint of danger, like the shadow of a passing bird or biologist's net, they actively swim to depth. And when the danger passes, they bob right back up and take another breath. From the water surface, the adult mosquito, balancing precariously on the skin of its earlier self, emerges vertically, pulling legs and wings out from its still-submerged pupal husk. I would have thought it akin to taking your pants off, one leg at a time. But YouTube videos suggest otherwise: the legs pop out in pairs or all at once and the fly can immediately walk on water or take flight directly.

Most facets of mosquito life depend on surface tension. An aerator added to a birdbath, or a pumped waterfall or fountain in a garden pond, is enough surface disruption to stymie mosquitoes' plans.

Mosquitoes did just fine before birdbaths and garden ponds; they reproduce in any standing water, and some species survive in very polluted and/or stagnant water. Breathing air helps them survive inferior water quality. We provide many suitable habitats for mosquitoes: discarded buckets, tires, beer cans, even a water-filled hoof print will do. That tarp that you ignored on your woodpile all summer may well be the source of winged vampires. And a clogged gutter, from a mosquito's point of view, is a linear pond in conveniently close proximity to fresh food on the hoof or sneaker.

So, eliminating places for water to accumulate can reduce mosquito populations around your home. Perhaps you'll eliminate some other tensions from your life in the process!

SPRINGTAILS—TIGGERS OF THE INVERTEBRATE WORLD

As we leaned over the Colchester Bog boardwalk, a student asked, "What's that black stuff on the water?" I suggested gently poking it with a twig. This elicited the expected response: as though ejected from James Bond's Aston Martin, tiny black flecks scattered, landing inches away and on my student's hand.

Springtails, aka the Tiggers of the invertebrate world, are often seen bouncing out of footprints and depressions in snow; hence another moniker: "snow fleas." Although they have six legs and hop, they're not actually fleas. They're not even insects. Taxonomic revisions have alternately kicked them out of and accepted them back into the insect club for decades. Springtails, who, as far as we know, don't much care how they are classified, are now in a class of their own: Collembola.

The word "Collembola" relates to an intriguing structure called the collophore, and is derived from ancient Greek words *kólla*, meaning "glue," and *émbolon*, meaning "wedge." The collophore is a telescoping tube behind the rear legs that springtails can extend to reach any and all parts of their tiny bodies. The collophore serves as a grooming mechanism, and also allows a springtail to aim the direction of its leap.

At the end of the springtail's abdomen is what gives it its spring: the forcula, which as the name implies is a forked structure, folds under the springtail's body like a jackknife and is held in place by a catch called a tenaculum. When faced with a predator (or poking twig) the hydraulically pressurized forcula is released, propelling the springtail up to three hundred body lengths away. Operating at our human scale, they'd comfortably clear anything on the Manhattan skyline.

Another common trait among springtails is a cuticle, or hard outer layer, that repels water. This is useful in melting snow, and it's also helpful for aquatic species like *Podura aquatica*, the springtails my student and I encountered at the Colchester Bog. This springtail's cuticle allows it to forage on still waters without drowning.

Podura aquatica lives on temperate water bodies throughout North America, Europe, and Asia and grazes on diatoms, plankton, unicellular algae, and rotting vegetation trapped in the surface film. These aquatic springtails have larger and flatter forculas than those of their dryland cousins, facilitating leaps from bogs and ponds without breaking the surface tension. When I watch them bounce from Colchester Bog, I don't even perceive the slightest ripple.

Bouncing is not the only way *Podura aquatica* gets around. In a 1915 thesis, George H. Childs, a University of Minnesota entomologist, observed they jumped only when disturbed and mostly got around by walking on the water surface. Jumping is risky when there's little flight control, and an errant springtail could end up high, dry, and away from a suitably wet habitat. Curiously, Childs also noticed that this species uses its antennae as a fourth pair of legs. In fact, when he removed the antennae, the springtails struggled to walk at all. Childs also discovered that springtails overwinter buried in frozen mud above the waterline.

There are some nine thousand springtail species worldwide. Most are landlubbers that occasionally end up on water, but *Podura aquatica* is not unique in its aquatic habits. *An Introduction to the Aquatic Insects of North America* lists eight springtail families specifically associated with water. Many of these live in the ocean's tidal zones and are best discovered by simply flipping over seaweed. But others, like *Podura aquatica*, are strictly fans of freshwater surfaces.

Felipe Soto-Adames and Rosanna Giordano of the University of Vermont studied springtails in the mid-2000s and included Lake Champlain in their collecting trips. They found a species entirely new to science near the Pine Street Barge Canal in Burlington. Fittingly, they named it *Scutisotoma champi*, after Champ, Lake Champlain's legendary monster. They named a second springtail, *Ballistura rossi*, for a much-loved colleague, Ross Bell. This species was found only on a constructed wetland at the University of Vermont, where Dr. Bell spent most of his career. A third species was named *Subisitoma joycei* for Joyce Bell, Ross Bell's partner both in life and in science.

It goes to show there is always something new to discover, some new science to invent, some frontiers that remain unexplored. The pond you explore in your neighborhood likely contains life-forms unknown to science, and maybe some of them are springtails.

SUMMER SKATERS

Scanning a sunlit pond floor for crayfish, I was distracted by seven dark spots gliding in a tight formation. Six crisp oval shadows surrounded a faint, less distinct silhouette. The shapes slid slowly and then, with a rapid motion, accelerated before slowing to another glide. I can remember seeing this pattern as a child, in my first explorations of pond life.

Water strider shadows are far larger than the insects casting them. To visualize the surprising proportion of legs to body, it may help to think in human scale. For mathematical simplicity, picture a six-foot-tall man lying flat on the water surface. Imagine that attached near his hips he has a pair of seven-foot-long, stick-skinny legs pointing back at a 45-degree angle. Just forward of these spindles he has another pair, these nine feet long and pointing forward at a 45-degree angle. A pair of three-foot-long arms point forward, and each has a single claw protruding from the palm.

The legs are long for good reason; they distribute body weight over a wide area and, aided by water-repellent hairs, allow the insect to coast across the water's surface tension. The minute leg hairs are densely packed, and each has many air-trapping surface grooves. According to

Lei Jiang and Vefeng Gao of the Institute of Chemistry in Beijing, who discovered the grooves, water striders displace enough water to float up to fifteen times their own body weight. This extreme buoyancy is enough to keep the water strider's body high and dry above the water, even during rainfall and choppy conditions.

Because these insects literally walk on water, some call them "Jesus bugs." When fish or backswimmers approach, the water striders are well positioned to make an aerial getaway. Their super buoyancy means that they can use their long legs to jump straight up from the water surface, and once airborne, they can spread their wings—yes, they have wings—and fly to safer haunts.

Slow-motion video reveals how water striders move. The longer middle legs sweep back rapidly like oars, pushing against the surface tension to drive the insect forward. Human rowers lift their oars out of the water on the recovery stroke to reduce drag, and rapidly moving water striders do the same thing. However, when moving more slowly, they drag their middle legs forward along the water surface. The rear legs trail and change angles like twin rudders steering the insect toward food or mates or away from hazards.

All the while, the front legs rest on the water surface just forward of the insect's head. Theirs is a murderous function, one allowing the water strider to find and seize its next meal. Subtle ripples made by surfacing aquatic insects including mosquito larvae or struggling terrestrial insects on the water surface function like tugs on a spiderweb, leading the water strider to its prey. The single-clawed forelegs grapple the prey while the insect's piercing mouthparts stab through the cuticle, consuming bodily fluids as if through a drinking straw.

To see this firsthand, my Saint Michael's College students and I dropped a few large carpenter ants onto the water surface of some ponds in Winooski. It took only seconds for a water strider to grab the first ant. Others were rapidly scooped up and carried off. A braver student dunked a yellow jacket, trapping her in the surface tension. The water striders investigated but took a pass on that risky meal. The yellow jacket climbed out on some vegetation a little the worse for wear.

My students and I were also curious to see if the insects were faithful to particular pools or if they moved around. We used paper correction fluid (Wite-Out) to mark a dozen water striders and released them where we caught them. The following day, we found marked water striders in their home pool, but also in pools upstream and downstream. We frequently observed water striders fighting each other. Perhaps territoriality and competition drive them to seek other living space?

FISHING SPIDERS

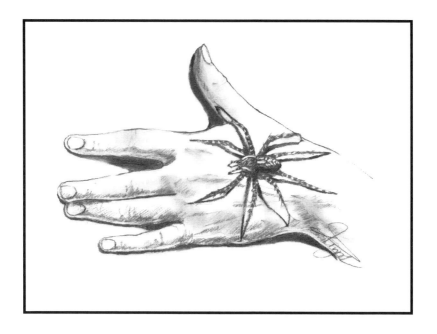

Large fishing spiders walking on water can be fascinating—or terrifyingly unnerving. The latter reaction is common among Saint Michael's College students as we sample Vermont's streams and ponds. On one occasion, a normally macho student screamed, dropped his net, and leaped from the stream to avoid a particularly large specimen. But have no fear; these beautiful beasts will not carry your offspring away. In fact, they are completely harmless—at least to humans.

Fishing spiders boast a leg span of up to three inches. Their rear legs cling to vegetation along pond and stream edges, while their front legs rest on the water's surface, awaiting the telltale vibrations that signal a meal—or a mate—is nearby. On class trips, we regularly encounter the appropriately named dark fishing spider (*Dolomedes tenebrosus*) on streams, and the six-spotted fishing spider (*D. triton*) more commonly on ponds. A third species, the striped fishing spider (*D. scriptus*), is also common in eastern North America.

Although these spiders can subdue small fish, they rarely do, opting more often for smaller prey. Manfred Zimmermann and John Spence from the University of Alberta recorded 625 fishing spider meals that included not a single fish, and only one tadpole. Water striders topped the menu, followed by adult damselflies, then aquatic insects (including mosquito larvae) surfacing for air. Not much farther down the fishing spider menu were other fishing spiders.

Both male and female fishing spiders will consume their own kind, although females—with a leg span of three inches—are twice as big as males, so are more likely the diner than the meal. Males have many behaviors to communicate "suitor, not snack." A male fishing spider seeks a female's pheromone impregnated silk strands on the water surface. Once a suitable mate catches one of a male's eight eyes, he proceeds slowly with a series of leg waves, drums the water with his pedipalps (a pair of appendages that flank spiders' jaws) and makes larger jerking motions that ripple the surface.

Horst Bleckmann and Manfred Bender of Goethe University Frankfurt determined that male fishing spiders briefly produce vibrations of less than fifty beats per second. Conversely, insects fighting the deadly grip of surface tension produce faster vibrations for longer periods of time. These cues can spell the difference between life and death for amorous male spiders. Males sometimes miss other critical information, however. According to Zimmermann and Spence, males often approach already mated females uninterested in romance and more interested in their next meal. In these cases, the male spider may well become a snack, rather than a mate.

Few spiders scamper and sail on water—or dive beneath it—but fishing spiders are uniquely qualified for the task. Among ten spider families examined in one study, fishing spiders were the hairiest, and their hairs more effectively repelled water than those of the other spiders tested. This permits unique travel options: they dive to avoid predators, swim actively underwater, and can remain submerged for thirty minutes enrobed in a glistening shroud of air. Young spiders spin silk strands and "balloon" on the wind. In addition to rowing and running on water, fishing spiders raise their front legs and sail away, pushed along the water surface by summer breezes.

Although fishing spiders do not use webs for hunting, they use silk for other purposes. They spin drag lines to reduce their risk of being swept away by currents. Females make two distinct silks layers to protect their eggs: an inner layer consisting of large, loosely spun fibers and an outer layer of finer fibers densely woven to repel water. Female fishing spiders carry their egg sacs until the spiderlings hatch, at which time the mothers spin their first web. Like other members of their nursery-web spider family (*Pisauridae*), fishing spider mothers construct silken homes for their offspring and remain with them for some time. Nurseries are often built under structures we humans create near water, yielding a second common name of "dock spiders."

So, if you find yourself slightly spooked by a large spider coming to rest on your kayak, or in a cool corner of your basement, as sometimes happens, take heart that you have met one of the best mothers of the invertebrate world. If nothing else, this spectacular organism may well have consumed some mosquitoes, depriving them of the chance of consuming part of you!

.

PART III

FLOWING WATER

The care of rivers is not a question of rivers but of the human heart.
—SHŌZŌ TANAKA

Imagine living your life in a wind tunnel. If you stood and fought the constant force of the air, you'd quickly be exhausted. You'd learn to press up against the walls or the floor to reduce your profile and find creative ways to hang on.

Organisms living in rivers and streams have evolved and adapted to the rigors of water constantly flowing in one general direction—downstream. "Going with the flow" may sound like an idyllic life choice, but for river organisms it would mean exiting preferred habitat for an unpredictable future location. To avoid this fate, some organisms resist flow using use slim and flat morphology, hooks, claws, tethers, suction cups, and ballast. Others use behavior to seek out what slack water exists in their tumultuous environment.

The paybacks for all of these specialized adaptations include a more abundant supply of oxygen than other freshwater habitats, reliable waste removal, and a conveyer belt of food at least for some. And river organisms from filter-feeding carp to caddisflies have adapted to consume the minute organic particles suspended in the water column by the constantly moving water.

In addition to rivers and streams, flowing water habitats include springs, seeps, and near vertical torrents where water cascades over rocky surfaces. While rocky substrates dominate steeper flowing water habitats, flatter, low-gradient rivers often flow over muddy riverbeds and these differences in substrate, water volume, bank width, and gradient explain why different life-forms dominate these different habitats. This part of the book explores the lives of some of these organisms.

RIVERS—LIFE IN FLOWING WATER

"You cannot step into the same river twice." Heraclitus's analogy of a river representing change in all things works precisely because rivers change dramatically from season to season, from place to place, and even from hour to hour. As I commute across the Winooski River bridge each morning, I register the river's dramatic responses to rainfall, snowmelt, or the failure of ice dams. A closer look reveals a less dramatic but constant redistribution of water, substrate, and organisms in space and time.

Every river, small or large, is its own agent of change. Flowing water carries materials in the journey downstream. From spring seeps to the world's largest rivers, "lotic" habitats are defined by flow; water enters these systems and exits with enough velocity to draw oxygen into the water, move substrates, and influence what life can persist there. Constant flow is, on the one hand, a conveyer belt delivering food, and on the other hand a potential force for destruction.

I recently enjoyed hiking on Mount Mansfield with my children. I couldn't resist the opportunity to turn stones in the mountain streams to see who crawled out. The small high-gradient streams were very different from the valley streams I typically sample. The water was crystal clear and the stream small enough that branches and twigs dammed its width, forming rocky step pools. Wet leaves piled abundantly and teemed with invertebrate life.

A New Orleans visit provided a dramatic contrast to the mountain stream. Behind the Audubon Aquarium of the Americas, I walked along the Mississippi just a hundred miles above the Gulf of Mexico. The river was a third of a mile across and cappuccino brown, colored by sediment and organic material washed from the landscape and buoyed by strong river currents. Hopping in to flip rocks was not a consideration.

Besides, rocks are uncommon on the muddy beds of large low-gradient rivers.

The journey from river source to ocean includes gradual transitions from very steep to very low gradient, narrow to very wide, shaded and leafy to sun-drenched, and constrained by small branches locked in place to whole trees moving with the flow. As tree shade decreases downstream, water temperature increases, causing oxygen concentrations to fall.

Physical changes cause biological responses. Almost one hundred years ago, European scientists developed the "Fish Zonation Concept." Fish like trout require cooler water with high oxygen content and live upstream. Moving downstream to warmer waters, trout are replaced by grayling, which in turn yield to barbell, followed by bream, who tolerate the warmest, least oxygenated water. This idea, but with different casts of fishy characters, applies to rivers on other continents.

In the 1970s, American scientists developed the "River Continuum Concept" describing changing macroinvertebrate communities from river source to mouth. Fallen leaves are the base of small-stream food webs and invertebrates called "shredders," including some species of crane fly (*Tipulidae*) larvae, which eat this tough material and dominate communities. Downstream, as water volume and stream width expand, the canopy opens and sunlight reaches the streambed. At this point, the food base shifts to include more "periphyton" such as algae that gives rocky surfaces their slimy coating. "Scrapers," or grazing invertebrates, consume this new bounty and become common. Finally, "filtering collectors" dominate larger rivers and dine on organic material that gives the Mississippi its mocha color and prevents light from supporting periphyton growth.

In addition to changing along the length of the journey to the sea, river organisms also experience dramatic seasonal changes. In the 1990s, I attempted year-round sampling in Vermont streams. Summer and fall samplings were easily accomplished. But winter brought impenetrable ice, and spring melt swelled the streams to hazardous levels. I abandoned the effort, but the attempt powerfully illustrated seasonal changes in streams. Fall establishes the food base of accumulated leaves. Although winter ice thickly caps many streams, water flows beneath, and invertebrates thrive and fatten on leaves left from fall. Spring melt brings floods, and invertebrates hunker down in nooks and crannies deep below riverbeds in the area called the "hyporheic zone." In summer, each insect species takes flight in a brief synchronized "hatch" to mate and lay eggs. These hatches happen from late spring to early fall with a few hardy species emerging in winter.

Streams also change rapidly in response to rainfall and snow melt. We installed automated samplers on Vermont streams to sample water as rivers swelled during storms. Small streams rose and fell quickly following rain showers while larger rivers receiving input from multiple tributaries responded more gradually.

Stormwater lifts and carries sediment that can sandblast most algae from river rocks, dramatically lowering the food base. Just the few algal cells remaining "reseed" and the algae regrow rapidly. River macroinvertebrates and fish are also well adapted to storm pulses. Plant and animal communities in other ecosystems are regulated by competition, predation, and herbivory. Major interactions occur between species. But river communities are more frequently dominated by disturbances caused by passing storms that swell water flow.

So, as you consider the organisms in this part of the book, keep in mind that their diverse adaptations have been fine-tuned by eons of life in hyper changeable disturbance-driven environments. Many of these organisms include blackflies (*Simuliidae*) and net-spinning caddisflies (*Hydropsychidae*) require flowing water for door-to-door food delivery. If these organisms could not handle gully-washer storms, they would have gone extinct. You and I, on the other hand, are not so well adapted to river life, so do be careful as you turn stones in your local streams.

.

NETS, BOOTS,
AND ICE CUBE TRAYS

I have a pre-pandemic memory of a dozen high school students—
armed with dipnets and wearing chest waders—emerging from a
Saint Michael's College van.

Before masks and social distancing, my collaborators and I packed
vans with students from Vermont, Massachusetts, and as far afield as
Puerto Rico to monitor macroinvertebrates in Vermont streams.

Because they are plentiful and respond quickly to environmental
change, macroinvertebrates are great indicators of river and stream
health. Beyond indicating the health of a stream, however, macroinvertebrates are themselves essential parts of intact stream ecosystems.
While some consider fish the most important animals in a stream, any
successful angler will tell you that without macroinvertebrates, there
can be no fish. Anglers depend on their quarry's appetite for invertebrates, whether they use crayfish or hellgrammites to lure bass, worms
for many fish species, or delicately cast flies for trout. Invertebrates are

the links between the leafy base of the stream food web and the fish that feed herons, kingfishers, and people.

Macroinvertebrates are, simply put, invertebrates large enough to be seen without magnification. To keep sampling consistent between fifty-five-year-old eyes and those of my young collaborators, we adopted the Vermont Department of Environmental Conservation's standard: we washed all of the debris collected in our nets through 0.6 mm sieves; anything caught on the sieve was part of our sample.

There are more than 1,500 recorded species of caddisflies alone in North America. These caddisflies and other insect and noninsect macroinvertebrates range in sensitivity to pollutants—from the larvae of some nonbiting midge species called "bloodworms," which dominate the nastiest of streams where most oxygen has been removed, to delicate stoneflies, mayflies, and other nonbiting midges found only in the cleanest streams.

Researchers use macroinvertebrates to indicate overall river health because these multi-legged citizens of the deep are there year-round and provide evidence of many potential impacts over their life cycles.

Eroded soil from construction sites, riverbanks, or elsewhere clogging the nooks and crannies where invertebrates live is a common impact that eliminates many species. This erosion is so common in our cities and towns that scientists call it "urban stream syndrome." But the absence of sensitive invertebrate species may also indicate less obvious water quality issues, such as a chemical spill long since washed downstream.

Scientists sample many local streams to measure what "typical" looks like in that specific area. Then they can compare a site of interest to the "reference conditions" averaged across several local sites. When streams are regularly monitored, a drop in macroinvertebrate diversity can be the first signal of water quality problems that merit additional investigation.

Sampling macroinvertebrates requires only a net and some inexpensive equipment. Importantly, researchers can garner high-quality data just by counting the number of "types"—or families—they find, even without noting specific species. A greater number of types indicates cleaner streams. For example, water samples from Snipe Island Brook, a woodland stream in Richmond, Vermont, generally contain about forty macroinvertebrate families. Potash Brook in South Burlington drains a mixture of woodland, urban areas, and a farm and typically yields twenty-six families. And Burlington's Centennial Brook, which drains several paved parking lots and an airport runway before entering the woods, hosts about a dozen families.

With patience, a microscope, and taxonomic keys, researchers identify macroinvertebrates to genus or even species. With only a magnifying glass and field guide, however, it is possible for any community scientist to quickly determine the macroinvertebrate's order, and often its family. The best part of this approach is that you can sort your bugs in an ice cube tray in the shade and then return them alive to the water. It is very important not to move them between streams, because introducing a new species may upset the local balance or put an endangered species at risk.

When sampling macroinvertebrates, it is important to watch the weather. Rain-swollen streams can be unsafe to sample, and many macroinvertebrates will have gone deep into the streambed to avoid the churning power of the flood. It's best to allow some time for them to come back up within range of your net.

I'm looking forward to getting back into Vermont streams this year with excited students. We'll do it safely, both from the point of view of water safety, and avoiding communicable diseases.

THE AFTERLIFE OF LEAVES

My sister Valerie and her husband recently visited with hopes of experiencing a New England fall. In preparation for the best leaf viewing, we exchanged weather forecasts and studied leaf maps. Nightly news showed peak leaf color perform its annual slow-motion march from the mountains down into the valleys, from northern New England southward. We hoped to time their visit to best take in a picture-perfect fall day reflected in ponds and lakes.

But once leaves fall, lakes, ponds, and rivers have their own leaf season. Many of the leaves that blow across the landscape accumulate in water bodies. Dry leaves hit watery surfaces, stick, and then eventually sink. The accumulated leafy piles provide most of the food base for everything in freshwater ecosystems, including bacteria and fungi, insects and other invertebrates, and fish. Ultimately, the nutrition from leaves works its way up through the food web to birds, otters, mink, bears, and even people.

I would imagine that few readers would ever consider casting a hooked leaf to catch fish, and with good reason: few fish eat leaves.

Several links in a complex food web are necessary before nutrients and calories from leaves ever feed a fish, which may in turn end up as a meal for a bald eagle or osprey, or on the menu at a fine Connecticut restaurant.

Before exploring the afterlife of leaves underwater, or even their importance as tourist eye candy, it is worth considering them from a tree's viewpoint. First and foremost, leaves produce food for trees. These little green power plants suck carbon atoms from the air and use the sun's energy to string them together to form complex carbohydrates—sugars, starches, and even the tree's wood fibers. When we consume maple sugar, or a caterpillar eats a leaf, digestion liberates the sun's energy, breaking down plant products to provide nutrition.

But providing leafy nutrition for herbivores is not in a tree's best interest. Hungry insects, such as spongy moths, can kill trees, and the trees are not going down without a fight. In addition to more obvious defenses like thorns and hairs that make leaves harder to get or consume, trees have more sinister weapons in their arsenal: they wage chemical warfare against herbivores bent on eating them.

Plants produce a dizzying array of indigestible and even toxic chemicals to dissuade would-be leaf eaters. We take advantage of this pharmacological bounty as inspiration for everything from aspirin to caffeine to cancer-fighting compounds.

Although the chlorophyll that makes leaves green breaks down to reveal spectacular fall colors, many of the unpalatable, indigestible, and poisonous compounds are far more stable and remain in falling leaves. So, when leaves first fall into lakes and streams, not only are they inedible to fish, they are of little nutritional value and sometimes still toxic.

A physical, chemical, and biological process known as "conditioning" converts inedible, leathery leaves into delectable snacks that underpin freshwater food pyramids. Leaves in streams and ponds leach their chemical compounds into water in much the same way that tea leaves release caffeine, flavors, and the tannins that color our drinks. But the process of leaching toxins and tannins from leaves in water bodies takes place over weeks, and in some cases months.

Releasing toxins is just the first step; there's still the issue of indigestibility. Most of a leaf's structure is cellulose, like the paper upon which this book is printed or the wood framing that supports a house. Animals lack digestive enzymes to break down cellulose. Even termites, perhaps the most infamous wood eaters, depend on microorganisms to glean nutritional value from your floor joists.

Two groups of freshwater microorganisms—fungi and bacteria—colonize sunken leaves and begin the digestive process before any animal takes a nibble. And still, even after the leaves are appropriately conditioned, most self-respecting fish will turn their noses up at the leafy salad bar.

But invertebrates—insects, crayfish, aquatic sow bugs, and many others—treat all of that leafy goodness as a smorgasbord tossed to perfection with fungal and bacterial dressing. Invertebrates called "shredders" make their living by munching through the piled leaves, consuming fungi and bacteria along with the leaves. Only by eating these invertebrates can fish finally access the enormous nutritional value that our fall leaves provide to freshwater communities.

Just as each leaf has visible characteristics unique to its species, leaf chemistry, physical structure, and the timing of leaf fall also differ from one species to the next. As a doctoral student at the University of Connecticut, Yingying Xie, now a visiting assistant professor at Northwestern University, used time-lapse digital cameras to show the end of leaf season in maples occurs more than twelve days before oaks. Dr. Xie's research certainly matches up with my experience in my own yard. It seems the oaks are watching and spitefully waiting until I've finished raking maple leaves before dropping their ample supply of leaf litter, sometimes on top of early snow. And I see beech leaves hanging on well into winter.

Leaf structure and chemistry add additional variables to the mix. Linden leaves are soft and reach edible condition rapidly. Oak leaves, in addition to dropping late in the season, are leathery and loaded with tannins that take far longer to leach out. The result of all of this variability is that a diverse forest provides a reliable food supply that lasts through winter, spring, and tapers off into summer. In addition to supporting healthy terrestrial communities that better resist pest invasion, diverse forests also sustain freshwater communities.

Longtime New Englanders know that variations in rainfall, temperature, and storms shift the peak of fall leaf season from year to year. This begs the question: What might a warming climate do to leaf season? And further, how might it impact underwater communities? The obvious answer might be that warmer weather would simply delay the season, and indeed anyone who has observed fall trees while driving south from Vermont to Connecticut would likely agree. But there's more to climate change than just warming. We certainly have warmer summers, but we also have increased rainfall in the autumn, and specifically more intense fall storms throughout the Northeast. In another study, Dr. Xie modeled the impacts of predicted climate changes on twelve deciduous tree species in a range of locations. She predicted earlier leaf fall for some species due to summer heat stress and heavy rainfall, while later leaf fall for others due to a warmer climate. All of this will vary by location.

The University of Vermont's Brian Beckage documented uphill shifts in the distributions of trees, with fir trees contracting their mountaintop ranges while the seeds of broad-leafed species, such as maples, were growing farther uphill than they did in the 1960s. According to Alan Betts of Atmospheric Research, between the 1990 and the 2015 versions of USDA's plant hardiness zone maps, Connecticut shifted from zones 5 and 6 to the warmer zones 6 up north and 7 along the southern coast. The most recently updated map moves coastal towns like Mystic, Connecticut, from Zone 5 in 1990 to Zone 7a in 2023. To find Zone 7a in 1990 you would have had to travel about 300 miles south along the coast to Delaware. And while trees can shift local distributions uphill, it remains to be seen if trees can redistribute at the regional scale at a rate equaling the rate of climate warming.

Dr. Beckage and Dr. Xie have done some Herculean modeling to predict where a very complex set of parameters will drive our tree communities. And while their predictions are not necessarily tidy, perhaps they make the case for protecting diverse tree communities to spread the risk of drastic change across multiple species so that some future version of healthy forests can persist. Perhaps there's also a case to be made for "assisted migration"—planting trees some miles north of their current distributional limits.

Of one thing I am confident: so long as leaves fall into streams and ponds, the invertebrates in these water bodies will feast on the discarded food source. Will New England streams start to look like those of Pennsylvania or Virginia? That remains to be seen and depends on the actions we take to mitigate climate change.

Aquatic invertebrates have evolved over eons to take advantage of an abundant winter food supply. For many freshwater insects, most growth occurs in winter, even as their habitat is capped with ice. Liquid water flows throughout winter and provides aquatic insects with a reliably constant temperature that is not guaranteed to their dryland brethren.

This past spring, I visited a vernal pool and watched caddisflies crawling beneath the ice as they foraged on different patches of leaves. For insects in particular, the timing of winter feeding and growth is important. For most species, adults emerge to mate in summer when the weather warms their muscles enough for flight. Few insects of any type can muster the warmth to fly in winter, and so winter larval feeding and adult summer flights are a boon for aquatic species.

To get a glimpse into the winter communities in your nearest small stream, grab a handful of submerged leaves and drop them into a basin of water. You'll be amazed by the stoneflies, crane fly larvae, and other organisms that crawl out as you gradually remove the leaves. You can briefly establish a streamside macroinvertebrate zoo in a plastic ice cube tray to view your catch of the day. Take care to detain your guests just briefly and return them where you found them so they can continue feasting during the prolonged aquatic leaf season.

And as for my family's leaf-season adventure, the timing worked well. Before the visit was over, we were lucky to catch spectacular golden red vistas of maple, birch, and beech reflected in the New England lakes and ponds that these leaves would later feed.

SUBMERGED SILK SPINNERS

A small boy asked, "What's your favorite insect?" I answered without hesitation: "Caddisflies." But it's not the short-lived adults, while charming in their own hairy mothlike way, that capture my attention. My caddisfly predilection is reserved for the larval stages that last for most of the insect's one- or, less often, two- or three-year life spans. These larvae, like their caterpillar cousins, make and use silk in ways that fascinate me. Silk permits their use of a wide variety of freshwater habitats and food sources.

Consider the caddisflies of the family *Rhyacophilidae*. Their name translates to "river loving," and this preference appears to have served them well. They are found in fast-flowing streams where they spin silk ropes that anchor firmly to rocky surfaces, helping them to defy the pull of currents, and stay off some trout's dinner menu. Like ice climbers using crampons, they also have impressive claws that grow right out of their rear ends. Their anal claws and silk lines keep their bulging, segmented, Michelin Man bodies secured while they scramble about, eating insects including other caddisflies.

A more patient hunting approach is used by caddisflies in the genus *Nyctiophylax*. They find shallow grooves in stream rocks, and roof them over with stretched silk. There they wait, as a hunter might in a deer stand, until some insect takes its fatal step onto their gossamer tent. I can't tell if this is a more or less effective tactic than the search-and-destroy approach of their river-loving cousins. *Nyctiophylax* larvae are certainly slimmer, but both groups are successful predators.

Two common caddisfly families are distinguished by their complicated silk nets. Philoptamids sieve food particles from water using loose, finger-shaped nets under riverbed stones. Hydropsychid caddisflies stand their rigid-framed nets right up in the water flow to maximize their catch. You can often find their nets by searching sunlit riverbeds for shadows on otherwise smooth rocks. Because these nets filter particles from the water column in sufficient abundance they can improve water quality. My Saint Michael's College student researchers and I find that hydropsychids are particularly common in streams receiving particles eroded from urban and agricultural landscapes. The particles are an abundant source of digestible food and can support large populations of these resilient insects.

Case-building caddisflies are more commonly found in cleaner streams. These species may have been the inspiration for the "caddisfly" name. In Elizabethan England, traveling salesmen—"cadice men"—attached their wares to their clothing. Similarly, case-building larvae construct silk-lined, sleeping-bag shaped body coverings and ornament these with materials found in their habitat. Often, materials selected are specific to the type of caddisfly. Some use sand grains, others larger pebbles, and still others use sticks and leaves. There are even caddisflies that cover themselves in growing liverwort leaves, and one species that builds its case entirely from snail shells.

Case-building caddisflies use materials gathered in their immediate environments, which, of course, helps with camouflage. Some species are less choosy than others. I have seen *Neophylax* cases made primarily from tiny pieces of broken brick eroded from a construction site. More than one intrepid artist has provided the insects with semiprecious stones and sold their bedazzled cases as jewelry. I once kept some case-building caddisflies in an aquarium for a few months. I dropped in some wooden match sticks, along with natural aquatic vegetation. They incorporated the matches into log-cabin-style cases.

In addition to camouflaging the larvae, caddisfly cases directly protect larvae from predators. While larger predators may ingest the caddisfly, case and all, smaller predators can be deterred simply because the case won't fit in their mouths. In one study, fish were more likely to attack and eat caddisflies when larger stones were experimentally removed their cases. And it stretches the imagination to consider how fish could ever hope to eat *Neophylax* pupae, entirely encased in stone, and fastened tight with silk to stream rocks and to each other in large numbers.

All of this silk production—for tethering and camouflage—is costly to the insects. By some estimates, caddisflies invest more than 10 percent of their resources in silk. This is a significant expenditure, and some caddisflies recycle silk by eating it.

Speaking of consumption, I'm reminded of Aldo Leopold's comment that one should seek to cultivate a "refined taste in natural objects." For whatever reason, my "refined taste" as an entomologist has developed as a preference for a silk-spinning, case-building aquatic insect. Of all the insects on the planet, why caddisflies? Why not? I hope no one asks me to choose a favorite caddisfly!

BRAINWASHED BY WORMS

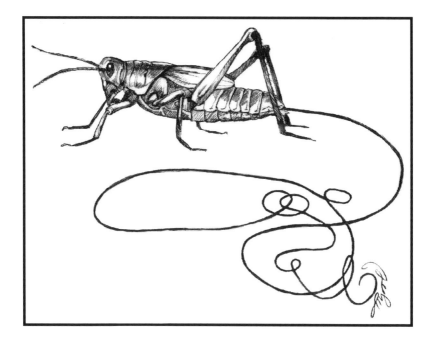

Some of my favorite children's books describe life cycles as heroic tales of persistence and redemption. From *The Ugly Duckling* to *The Very Hungry Caterpillar* to *A Seed is Sleeping*, these stories have brought the miracles of growth and maturation to life for generations of readers. I can't say, however, that I know of a single children's book that describes the impressive hero's journey of Nematomorpha, commonly known as horsehair worms.

These curious creatures earned their name because they are sometimes found in watering troughs and were once believed to be horsehairs that became animated. Many insects might well be very pleased if this were true.

The improbable details of the horsehair worm's life cycles are as convoluted as they are gruesome. Horsehair worms begin life as strands of gelatinous eggs in aquatic habitats including streams, lakes, puddles—and, yes, horse troughs. After hatching, the larvae, scarcely as large around as the thickness of a page, remain on pond and river floors until they are eaten by something else. While creatures from fish to stoneflies accidentally ingest these larvae, not all hosts are equal when it comes to horsehair worm survival.

Once inside this first host, the larva burrows through the digestive tract into the host organism's body and forms a resting cyst. Two stars must align for success of the cyst's next life stage. First, the host must be in an insect; if the cyst ends up in a fish, snail, crayfish, or some other noninsect being, it is doomed. Second, the cyst must survive the host insect's immune defenses; some hosts wrap the cysts in compounds that can kill them.

Should a cyst survive, the next leg of the journey takes it out of the water. Aquatic insects emerge from water to reproduce, and horsehair worm cysts stow away for the trip. Exiting the water comes with its own hazards, for both insect and passenger; hungry fish may well dine on the paired travelers, at which point the horsehair worm cyst is just a fish-food morsel. Many female aquatic insects return to water to lay eggs, a journey that can be as hazardous as was the exit. For a horsehair worm, a host insect's successful return to water spells death.

Many aquatic insects, however, live their final hours on dry land, where their bodies eventually become scavenger fodder. This is where the horsehair worm's path takes a more gruesome turn. Once again, not just any scavenger will do. Horsehair worms favor specific hosts that seem particularly susceptible to the worm's brand of mind control, including crickets, cockroaches, and beetles. There's nothing a cyst in a dead mayfly, caddisfly, or dragonfly can do to ensure it is eaten by a favored host. But when cysts are consumed by the right insect, they set about coopting the host's resources—and eventually its free will.

Ingested cysts enter a distinctly wormlike life stage and grow rapidly, nearly filling their host's abdominal cavity. After reaching their full size, the worms release chemical messengers that seem to scream "dive and swim." Otherwise, sensible terrestrial insects that would have spent their entire lives on dry land are suddenly possessed by an irrepressible urge to swim like Michael Phelps.

Once their hosts are dunked, the worms emerge into the water via holes they poke through the host's body wall. Uncoiled adult worms can measure four inches in length and resemble a fine leather shoelace. Once in the water, they mate and lay their eggs. Surprisingly, some host crickets and cockroaches survive this bizarre infestation, scramble ashore, and go on to feed and even reproduce.

One might ask how a simple worm could possibly survive plot twists worthy of a Tolkien tale and then repeat the process generation after generation without interruption. The answer comes down to a common strategy among parasites: strength in very large numbers. Horsehair worms lay millions of eggs, and the larvae are often the most common life-form on the floors of aquatic habitats. By overwhelming the odds with extraordinarily large numbers, horsehair worms successfully produce enough offspring to keep these fascinatingly macabre animals around.

And so, if you happen to glimpse what appears to be a swimming pencil lead, or perhaps a Gordian knotlike tangle of such worms, in your dog's dish—or, more alarmingly, in your toilet bowl (it happens)—have no fear. Just look around for a rather damp and slightly confused cricket. What a tale they'd have to tell!

IF IT LOOKS LIKE A SNAIL, IT MIGHT BE A CADDISFLY

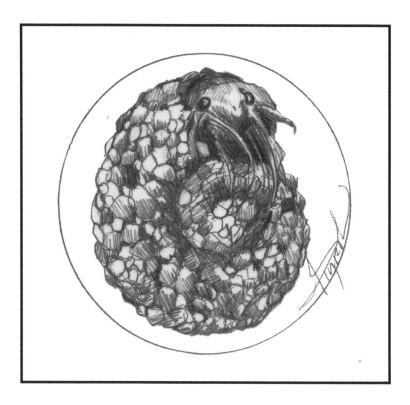

While sampling in the LaPlatte River, students noticed what looked like rough black pebbles about the size and shape of well-worn pencil erasers. I suppressed my mild distress as they started to discard the "pebbles." When sampling aquatic insects, I discard little.

I gathered the students around and balanced one of the pebbles on my finger and simply said "watch." Shortly thereafter, the pebble was making its way off my fingertip seeking wetter and cooler places.

We placed several of the animals in dishes and distributed them along with hand lenses. Discoveries and observations were made on all sides. "When you see it up close, it's obviously a snail." "Hang on, it seems to have legs." "It just blew a bubble out the back end." "Hey, they pull back into the shell if you poke them." "Somehow they stick to my forceps."

Unplanned observations testing no particular hypothesis are essential components of field biology. The animals were snail-cased caddisflies, *Helicopsyche borealis*. When empty cases were first observed, a biologist described them as snails with the unique ability to incorporate sand grains into their shells.

The truth is more interesting. Caddisflies in the genus *Helicopsyche* use silk to bind sand grains together to make protective cases. Case making is common in caddisflies, but snail-shaped cases are unique to just one genus, at least in North America. That snails and caddisflies have evolved to produce very similar protective structures is a remarkable example of convergent evolution.

In common with snail species, *Helicopsyche* shells, or cases, generally coil in one direction. When Robert Hinchliffe and Richard Palmer from the University of Alberta examined 150 *Helicopsyche* cases from the Royal Ontario Museum collection, they found that all coiled to the right. When I look at preserved specimens in my collection, I have yet to see any coiled to the left, but now I'm inspired to look more carefully.

As my students looked carefully at their specimens, they made observations I can't make from pickled samples; for example, the bubble that escaped from one case. Like many caddisflies, *Helicopsyche* larvae wriggle to create small water currents through their cases. This brings in oxygenated water and flushes away caddisfly waste. The waste produced by a single *Helicopsyche* must be scant, indeed; but in aggregate I'm sure that their output, together with that of other species grazing on the rocks of the LaPlatte River, provides organic matter for an array of other tiny organisms that eke out their existence by gathering or filtering particles from the water.

Helicopsyche larvae eat periphyton—a blanket term that refers to the algae, diatoms, bacteria, and fungi that grow on clean rock surfaces. In some situations, the snail-shaped *Helicopsyche* case serves as a mobile garden. Jennifer Cavanaugh and colleagues from the University of Wisconsin discovered that larvae from one caddisfly species grazed periphyton right off the cases of their peers. It would be interesting to see if *Helicopsyche* cases provide similar movable feasts in the LaPlatte River.

Why the caddisfly cases stuck to students' metal forceps proved to be the most challenging question of the afternoon. My first thought was that the caddisflies were grabbing on with their six limbs, as river insects do. My students rapidly disabused me of this simplistic notion. I was schooled when one student showed me that even when the insects were completely withdrawn into their protective armor, the cases seemed magnetically attracted to the forceps.

Magnetism proved to be the answer. After preserving some *Helicopsyche* cases in alcohol, and bleaching off the periphyton, the mystery was solved. The now-empty cases still clung to metal. On examination under the microscope, small black grains were present among the brown and white sand grains. And when I carefully disassembled the cases, and dissected out the black grains, only the black grains were attracted to the forceps. The caddisflies had incorporated some magnetite into their cases.

We counted our caddisflies and other insects before returning them to the stream. Data sheets were filled, metrics of biodiversity were calculated, and hypotheses were tested; this was the planned point of the trip. But a great many more interesting lessons were learned simply by watching an insect as small as a shirt collar button. I like that it was many people looking closely and that each person noticed something different—the best part of an overall lovely afternoon, I thought.

FLAT AS A PANCAKE

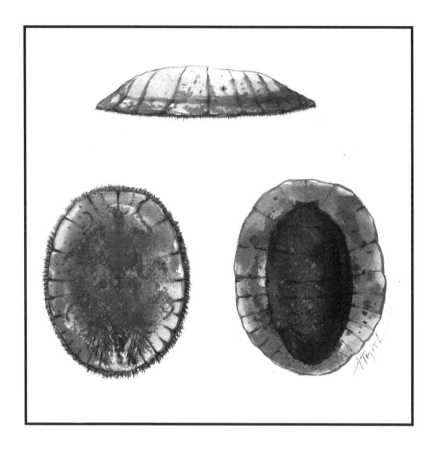

Imagine for a moment that you travel on all fours like other self-respecting quadrupeds. Extend your imagination a little more, and with it your body, so that a large dome-shaped shell-like structure extends out to cover you in all directions. From above, a predator would see only a disk with a snug fit to the ground on all sides. Now shrink dramatically and move into the nearest fast-flowing stream: you are well on your way to becoming a water penny beetle larva.

If you consider the characteristics just described as distinctly non-beetle like, you are not alone. The same notion struck the nineteenth-century biologist who first peeled one off a rock and sat back to classify it. His confusion created a minor kerfuffle in the scientific literature that leaves a lasting legacy.

As the name suggests, water pennies are about the diameter of a US or Canadian penny. But diameter and near circular shape is where the analogy ends. Water pennies are not nearly as thick as a penny, and this flatness allows for their survival in a fairly hostile environment.

Water pennies are found on rocks in flowing water. But the flow that discourages silt from fouling their habitat also presents the constant risk of being washed downstream. Luckily for water pennies, flow is not the same everywhere, even in the fastest-flowing stream.

The deepest and fastest part of a river or stream is called the thalweg. Often, it's in the center, but rivers turn, loop, and meander, so the thalweg's exact position is unpredictable and shifts over time. The fastest water in the thalweg is nearest the surface, and velocity reduces with depth.

Evolution has prepared water pennies very well for the hazards of flowing water. Getting away from the thalweg, riverbed friction slows water velocity dramatically the closer one gets to the rocky riverbed. In fact, on riverbed rocks there exists a thin layer of water with virtually zero water movement. This peaceful haven just below the chaos is called the "boundary layer."

A water penny's low profile allows it to live almost entirely in the still-water boundary layer. Because the edge of a water penny's disk slopes down toward the rocky surface, water flowing over the larva would tend to exert some downward pressure even when the beetle ventures to more exposed rocky surfaces. The larvae eke out a living by grazing on periphyton, the complex mixture of algae, bacteria, and fungi that grows on rocky surfaces. Because of this combination of diet and habitat use, they thrive in flowing water.

These beetle larvae found a unique solution to surviving river conditions, oblivious to the confusion they might cause for scientists. A disklike shape certainly does not scream "insect," much less "beetle," and so when water pennies were first described by James Ellsworth DeKay in 1844, he did not place them with insects. In fact, he made the bold choice of placing *Fluvicola herricki*, as he named his newly described species, among the Isopoda along with pill bugs, roly-polies, and woodlice. It's not as though he miscounted the legs; in fact, he made it very clear in his description—"feet three pair." He was, however, at a distinct disadvantage since he had never seen an adult water penny. I sampled streams for nearly a decade before I saw one, so I forgive the esteemed Dr. DeKay his oversight. Besides, his was far from the only confusion.

Dr. John L. LeConte claimed the water pennies for the insects in 1849, naming them in honor of his father, with whom he happened to share a last name. And so, the new scientific name became *Eurypalpus lecontei*, but not for long. Samuel Stehman Haldeman changed the name to *Psephenus lecontei* in 1853 because the name "Eurypalpus" was already in use for a genus of true flies. (To make things even more confusing, Fluvicola—the original genus name—already applied to a group of birds.) These scientists would have loved a quick Google search.

Finally, Charles Leng and Andrew Mutchler in 1927 renamed the species *Psephenus herricki* because DeKay's species name took precedent over LeConte's; no disrespect intended to him or his father. This is the name we use today, and so if you peel a water penny off a rock, you can proudly tell your friends you have found *Psephenus herricki*, a true water penny! Unless of course you found a false water penny. But perhaps that is a story for another day?

SPRINGS IN WINTER

On a clear mid-winter day several years ago, my student Sarah Wakefield and I pulled on snowshoes, donned backpacks, and headed up through Smuggler's Notch in Vermont's Green Mountains. Our destination was Big Spring, which rises from Mount Mansfield's bedrock before flowing east for one hundred yards and then entering a culvert under Vermont State Route 108. When it emerges from the culvert, the spring water joins a stream fed by surface runoff and snowmelt.

Groundwater-fed springs can flow year-round, regardless of plunging temperatures that layer thick ice on surface water streams. Spring water maintains a constant temperature, due to its long-term contact far below the surface with bedrock, and the deep sands and gravels of subterranean aquifers that are protected from the vagaries of the weather. Springs take on the average annual air temperature of their locality, and Big Spring's water measures 41 degrees Fahrenheit in midwinter, just as it does in August. This constant temperature provides refuge for tiny creatures from both summer heat and the freezing cold of winter. Some species thrive only in the consistent conditions in springs, while others avoid them entirely.

Some spring-dwelling species are widespread—wherever there are springs. The cast of characters includes caddisfly larvae (Order Trichoptera): case-building, net-spinning, or free-living wormlike insects with six legs at the front end and a pair of hooks bringing up the rear. They eat leaves, algae, filtered particles, and other invertebrates. The caddisfly *Adicrophleps hitchcocki*, for example, can be found at most times of the year in its perfectly squared-off green case of living moss strands in springs all along the Appalachians. Other species are endemic to one or a handful of springs; the caddisfly *Apatania blacki* has been recorded from just two Pennsylvania springs.

Temperature is not the only difference between spring water and surface water. Surface water draining through leaves and soil carries particles of organic material that feed blackflies and other invertebrates equipped to filter tiny morsels from water. Net-spinning caddisflies are particularly adept at dining on this movable feast of water-borne particles. Springs, however, where bedrock and subterranean sands have filtered groundwater to remove these particles, tend to lack net-spinning caddisflies that are common in other streams.

But where there's a pattern, there tends also to be an exception, and one net-spinning caddisfly bucks the trend by specializing in springs. *Parapsyche apicalis* spins a far larger meshed net than do most of its cousins. This net is used more like a spiderweb than a filter, and *Parapsyche* larvae will eat any macroinvertebrate that stumbles across their threshold.

P. apicalis was one of the first macroinvertebrates that Sarah and I found during our winter expedition to Big Spring. It showed up in most samples along the length of the spring but disappeared abruptly at the confluence with the tributary below the culvert. Patterns like this are not unusual in spring brooks. In Pennsylvania's Mall Spring, four species found at the source disappeared downstream. And several other species occurred only downstream. Some combination of temperature and food sources likely drives these patterns.

Some theorize that rare spring-source species are "glacial relics," cold-water species left behind when cooler conditions moved north behind retreating glaciers from the last ice age. If this is true, it begs the question: What will happen to these species in a more rapidly warming planet? When Sarah and I visited Big Spring in the 1990s, a warming planet was less on our minds than was the need to stay warm and dry while collecting insects located under very cold water. We brought our pickled samples back to the lab for identification and quantification.

But there's no need for microscopes or vials of alcohol to glimpse the world of spring-dwelling organisms. If you happen upon a pool of flowing water emerging from the rocks in an otherwise frozen landscape, you may well have found a spring source. By lifting a rock from the streambed, you can see if anyone is home: debris stuck to a clean rock surface suggests a net-spinning caddisfly retreat. An inexpensive clip-on macro lens for your cell phone will allow a closer look. You can upload your photos to iNaturalist, and who knows? The community of naturalists may determine that you found *Adicrophleps hitchcocki* or another spring water specialist.

THE WINTER CADDISFLY

*Left: Male winter caddisfly (*Dolophilodes distinctus*). Right: Female winter caddisfly.*

On a March afternoon in 1994, I accompanied Professor Jan Sykora, my thesis advisor, on a field trip to the Carnegie Museum's Powdermill Nature Reserve in Pennsylvania's Laurel Highlands. Nearing our spring-fed research site, Jan removed his hat and tossed it on the snow. He procured forceps and a vial of alcohol from his pocket, lifted the hat, and picked up from under it a small dark insect, which he dropped in a labeled vial.

"*Dolophilodes distinctus*," he exclaimed, "male." I was impressed! To a new grad student, pulling the Latin name and gender of an insect out of a hat—especially when that insect was small enough to perch comfortably on an aspirin—seemed like a superpower.

Later, as Jan shared his wealth of knowledge, I realized how he had accomplished this hat trick. *D. distinctus*, the tiny black gold speckled-wing or winter caddisfly, has an unusual life cycle, and it's one of the few caddisflies that you're likely to find wandering around in the snow.

This species is a finger net web spinning caddisfly (family *Philopotamidae*), meaning that, instead of competing with many other aquatic invertebrates for relatively large food items, its larvae spin silk nets to catch very fine particles of organic matter drifting by in the stream water. Of the four caddisfly families that ply nets to gather particles,

Philopotamidae spin the finest mesh, with openings measured at approximately 0.5 by 5.5 microns; to put that in perspective, if you are reading this article in a hard copy of the magazine, about one hundred of these mesh openings would span the thickness of the page.

The larvae spin their nets on the undersides of stream rocks, anchoring them at the upstream end and using grooves in their silk-gland openings to simultaneously lay down seventy silk strands. Strands along the length are relatively stout, while those that run parallel to the upstream opening are finer. The nets are intricate structures, requiring in total more than half a mile of silk, but they're hard to see in all of their inflated glory. The moment you lift a rock from a streambed, they collapse in soggy piles. What you're more likely to see is a slimy patch on the underside of the rock, with a straw-colored larva wriggling its amber head out of one end.

Caddisfly species go through five larval instars. Mature fifth-instar larvae pupate, and a little while later, adults emerge in a synchronized "hatch." These mass events allow males and females of a species to find each other, and timing varies depending on water temperature. That is why hatches often start downstream, where the water warms sooner, and progress upstream. Most caddisflies have a single hatch in spring, summer, or fall that spreads over a few days to weeks.

The winter caddisfly, however, doesn't follow this pattern. Instead, it has two hatches, each extending to months, and one of which occurs in winter. This detail about timing was what Jan needed to identify the caddisfly under his hat. There are precious few winter-emerging caddisflies, and the size and color of *D. distinctus* is particular to that species. Seeing a small black caddisfly was all Jan needed for identification to species. A closer look would have revealed subtle buff to golden flecks on the wings that inspired its common name: "tiny black gold speckled-winged caddis." But how could he tell the gender of a ⅜" insect without so much as a hand lens? Again, the answer came back to caddisfly natural history.

When *D. distinctus* emerges in summer, males and females fly about in the usual way, and males follow pheromones, or chemical signals released by the females. But in winter, the females forgo wings entirely and wait, often on snow, for winged males to find them. So, the winged caddisfly that Jan lassoed with his cap could only be male.

Another aspect of the caddisfly genus *Dolophilodes* that is unusual: the pupa actually runs along the water surface to the stream edge. Caddisfly pupae are firmly attached to rocks beneath the water and use sharp mandibles to liberate themselves from their cocoon before swimming and scrambling ashore and finally shedding their pupal skin. In his book *Caddisflies: a Guide to Eastern Species for Anglers and Other Naturalists*, Thomas Ames describes particular techniques for mimicking *Dolophilodes'* unique sprinting pupa behavior to tempt hungry trout.

If you happen upon a hatch of these fascinating insects, reach for a surface fly such as a Palmer Caddis from your fly box. If you move it just so along the water surface, you may well be rewarded with a handsome brook trout.

THE SMALLEST ENGINEERS

Each year I pitch this question to my Saint Michael's College students: "What animal engineers its habitat?" It does not take long to extract an answer: "Beavers of course," and my students are correct. At a human scale, and in our part of the world where beavers have bounced back from past hunting pressures, this is the most obvious answer.

But life operates at many scales and the overwhelming majority of life happens at scales we largely overlook. Mosses construct vast bogs; minute corals made the Great Barrier Reef; and elephants famously dig for water quenching the thirst of entire savannah communities. Ecologists term these natural activities "ecosystem engineering."

When I picked a stone from the bed of the LaPlatte River in Charlotte, Vermont, one summer, I unwittingly uncovered a local ecosystem engineer that was significantly smaller than any beaver. I was studying seasonal shifts in macroinvertebrates throughout the year. One end of the stone I lifted from the streambed was thickly coated in caddisfly cases. My gut reaction was "This sample will be a nightmare."

Staring down a microscope at my first caddisfly larva was a revelation of nature's incredible complexity, a feat of fine-scale engineered beauty and a construction of silk and stone assembled following plans evolved over millennia. The pleasure taken from your first ten or one hundred macroinvertebrate samples depends on your particular passion, or perhaps tolerance, for bugs. But science is built on replication, repeated experiments, and observations. "Do it again" until you are certain. At some point, even the most patient scientist finds a limit.

And so this thickly "bejeweled" or perhaps "befouled" stone really tested my commitment to the science. The next rock over certainly hosted far fewer bugs and would get me more rapidly to the requisite level of replication for my study. I was tempted to defeat my carefully considered sample randomization procedure, discard the offending cobble, and "randomly" select a different, cleaner one. Yielding to temptation would have invalidated my study and so I dutifully scrubbed every last scrap of matter from the rock and preserved the lot in the name of posterity and science. And I was very glad I did!

Back in the lab, I postponed looking at the dreaded sample and eventually convinced a student collaborator to share the work. As we delved through the organic matter lovingly referred to as "crud" to extract invertebrates, some fascinating observations emerged that suggested even more questions and eventually an entirely new experiment.

Most obvious among the pickled beasts of course, were the caddisflies. They turned out to be *Neophylax fuscus*, all eighty-seven of them from this one chunk of rock no larger than a fist. Caddisflies in the genus *Neophylax* range from North America to Japan and eastern Russia. The larvae pile themselves up densely on rocky streambed substrates before pupation. Some species rest in these piles all summer. The larvae fasten their sand and pebble-clad cases down with silk and unless they suffer the misfortune of being sampled by some Saint Michael's College student, they remain hunkered down between June and late September.

I suppose none should be surprised to find caddisflies in a caddisfly aggregation, but what else we found was the real story. After carefully setting the caddisflies aside in their little stone cases, we realized that more than three-quarters of the work remained yet ahead. Hidden among the silt, dead leaves, and algal strands were invertebrates, lots of invertebrates!

Living in the nooks and crannies between dozens of stony *Neophylax* cases, about three times as many other invertebrates had made their homes. We found *Oecetis avara*, a much smaller caddisfly that makes a cornucopia-shaped case of fine sand grains, midge larvae in great abundance, and many, many aquatic mites. One mite, in particular, *Torrenticola rufoalba*, seemed to favor empty caddisfly cases.

We learned about the mite's preference by gluing empty caddisfly cases to bricks and placing them in the stream for six weeks. Other bricks had sealed caddisfly cases affixed and yet another group of unadorned bricks served as controls. We found three times as many microscopic mites on sealed cases compared to controls; but with open caddisfly cases, mite abundance jumped to ten times the numbers sampled from control bricks.

If you look long and hard enough you too may find that there are tiny engineers tweaking the natural world in your neighborhood. To be sure, the effects may not be as spectacular or obvious as beaver dams, but they matter nonetheless. We know that mice move into abandoned bird nests; carpenter ants exploit tree holes made by woodpeckers; and our trail cameras at Saint Michael's College revealed that fallen trees serve as highways and bridges through vegetation for woodland mammals.

Does some organism use the silken homes made and later abandoned by tent caterpillars? Do spiderwebs serve some as yet undiscovered purpose once the arachnids no longer need them? There are far more questions in biology than there are answers, an exciting thought for any inquisitive scientist!

LIFE, DEATH, AND BLACKFLIES

I was in southern Connecticut one May to pick my son up from college. While he took his last exam, I took myself up a local hiking trail. Connecticut blackflies are as bad as their Vermont cousins, and I brushed several of the little beasts out from under my hairline. It can be hard to think of these biting flies with anything but disdain, but they do serve important ecological functions. And in at least one case, they also solved a murder.

Blackfly larvae are little, black, and shaped like bowling pins. They live in rivers and streams, and on their back end they have a circle of hooks that keeps them from being washed downstream. This is essential for survival; drifting organisms can rapidly become fish food.

Each larva attaches a clump of silk to a rock and then uses its hooks to get a firm grip on the silk. This leaves the other appendages free to collect food. "Appendages" is a loose term; like other true flies (order Diptera), blackfly larvae lack legs entirely. But where there's a need, nature provides a solution, and in the case of blackflies, their mouth parts have evolved into an elaborate pair of fans.

The fans open up like many-fingered hands with countless hairs attached to each finger. The open fans reach upward, grabbing particles that drift by and very effectively filter the water for morsels of food too obscure, too microscopically tiny, for the average human, or fish, to notice. When my Saint Michael's College students first find one in a preserved sample, I encourage them to gently squeeze the larva's head with forceps. Long-dead specimens will still open and close their fans right there, filtering the alcohol under the microscope. And in rivers and streams, the water is cleaner from all of this filtering.

Blackflies fatten underwater and emerge as little flies that crawl on us in search of blood. For many of us, the flies are a mere annoyance; for some they elicit allergic reactions. Many of the fattened larvae and emerging flies become fish food. The pupal stage, between larva and fly, occurs in silk structures that are fastened firmly to hard substrates and would be challenging for the average fish to dislodge. And it is the pupal stage that brings us back to the murder case.

In June 1989, scuba divers discovered an upside-down car in a deep hole in the Muskegon riverbed in Michigan. Inside lay the remains of a woman, well preserved by the cold river waters. Determining time of death proved difficult, but all evidence suggested foul play, and her husband quickly became the prime suspect. Among the evidence collected from the vehicle were three insect varieties, but it was the blackflies on the car's windshield that turned out to be the key witnesses.

In fact, the blackflies were long gone, leaving only their empty silk cases to suggest they had ever been there. Yet these small silken structures were all that was needed to break the case. With underwater crime scenes, detectives are tasked with determining the PMSI, or postmortem submersion interval. With a little knowledge of aquatic insect life histories, PMSI can often be determined, or at least narrowed down.

And so, scrapings from the car windshield and elsewhere were sent to Dr. Richard Merritt at Michigan State University, one of the world's foremost blackfly experts. The timeline of the particular case hinged upon a few facts: the husband claimed that his wife had left the house on a foggy September night following an argument. He also claimed to have heard from friends who had been in contact with her during the subsequent winter and spring. The blackflies offered a different opinion.

By examining a few scraps of silk secreted and assembled by black-flies on the windshield of a car, Dr. Merritt concluded that the car, discovered in June, had been submersed at least since fall of the previous year. This particular blackfly did not build cocoons during the cold of the Michigan winter, or spring; only during fall.

And so these obscure, and sometimes annoying little flies, were used to refute the husband's story of his wife whiling away the hours through winter and spring. Blackflies, along with other evidence, sent him to jail for a twenty-year sentence.

And although it is unlikely that the blackflies swarming around your head or mine will solve any crimes, perhaps we can take solace in them cleaning the water and providing the link between microscopic food particles and the fish we love to catch.

PART IV

LAKES, PONDS, AND OTHER STANDING WATERS

I will arise and go now, for always night and day
I hear lake water lapping with low sounds by the shore

—WILLIAM BUTLER YEATS

W. B. Yeats reminds us that standing water is not still water, and anyone who has experienced a storm on a lake would agree. As a small child, my family and I were caught unawares by a storm that turned the Shannon's Lough Ree into a force to be reckoned with. My father's wise solution was to make for the shelter of Hodson's Bay, and we walked home the long way around.

But movement in standing water helps answer the question of what is a lake? "Waves" would be the short answer, but enough wind-driven water movement to mix the layers of a water body generally defines a lake. Water mixing in ponds happens mostly because of convection driven by water temperature changes under the influence of air temperature and the sun.

The lack of unidirectional flow separates standing water from flowing water. Less flow means less erosion and more deposition. Nowhere is this more obvious than where rivers flow into lakes, and sand and silt suspended by river flow begins to settle out. Lake Champlain's shores show results of deposition in the form of river deltas. The Winooski and Missisquoi rivers both snake through deltas of their own making.

Deposition in standing water feeds lake-floor organisms adapted to these conditions and excludes riverine species that require regular flushing to keep their habitats tolerably clean. Lamp mussels, blood-worms, and burrowing mayflies are among the organisms that thrive in mucky lake floor sediments, often beyond the reach of daylight.

Other organisms such as fish and phantom midges stay neutrally buoyant using swim bladders and gas vacuoles to keep them up in the water column and above the muddy floor. This part of the book is devoted to these and other denizens of standing water.

LAKES—LIFE IN STANDING WATER

University of Vermont professor Ellen Marsden and I dove into Lake Champlain near Crown Point one afternoon to install zebra mussel sampling equipment. As we ventured deeper, visibility declined until all I could see was Ellen's fluorescent scuba tank, and I was very glad of the anchor line we used to guide us to our target. While Lake Champlain is murkier than some lakes, light declines with depth in all lakes.

Even the cleanest water absorbs light, but plankton, floating plants, algae, and sediment washed in or resuspended from lake floors, all block their share of light. Light traveling for some distance below the surface supports enough photosynthesis to meet the oxygen needs of fish, invertebrates, and plankton. But as we travel deeper and darker, a point is reached when the sum of the lake organisms, including plants and algae, consume more oxygen than photosynthesis produces.

The shallow, brighter, well-oxygenated water called the "photic zone" extends all the way to the floors of shallow lakes and ponds. But in deeper lakes, water extends well past the reach of sunlight into the "aphotic zone," defined as receiving less than 1 percent of the light that strikes the water surface.

I was interested to see how Lake Champlain's photic zone measured up to my observations while diving. Steve Cluett, the captain of the University of Vermont's research vessel, shared some numbers he gathered by lowering a light-measuring probe into the lake. Compared to light at the surface, light penetrating Lake Champlain fell below 1 percent between forty and sixty feet deep depending on day and location. My dive was shallower than thirty feet, and I could still see, so my observations matched up fairly well with Steve's data.

Temperature also varies with depth in lakes, and in summer we expect warm, less dense water to float over colder, denser water. This layer cake pattern changes through the seasons as described in the essays in the first part of the book. The organisms of standing water handle temperature and light differences in addition to other physical and biological constraints that are very different from those in flowing waters. Planktonic organisms, for example, have huge populations in lakes and ponds but are rare or absent from most streams.

Standing waters in lakes and ponds are termed "lentic." Without constant flow, sediments settle to lake and pond floors and accumulate. Decomposing organic material consumes oxygen making it difficult for most organisms to survive on lake floors. However, an array of hardy organisms not only survive, but thrive right on the mucky surface, and still others have adapted to life directly in the muck.

Bloodworms (*Chironomidae*), larvae of some nonbiting midge species, are one such muck-adapted group of organisms that my students and I sample from deep in Lake Champlain. A form of hemoglobin bound to their cuticle gives them their red color and common name. They live in shallow tunnels in the mud and wriggle their bodies to draw water into the sediment where their hemoglobin absorbs oxygen allowing them to live in this oxygen-starved environment.

Despite the oxygen shortage in the deep, there is a bright side in this darkest of environments. The organic content of the constant "rain" of material from above is a reliably delivered food source to a place that lacks the light to grow its own.

Were one to start from a dark lake floor and walk uphill toward the shallows and the light, eventually you'd find a place where the lake floor does actually grow its own food. Once you find that rooted lake-floor vegetation, you will have entered the "littoral zone" and left the deeper "limnetic zone" behind. All of the submerged solid ground you walked upon during your lake trek, dark or otherwise, or any solid ground under any water body is known as the "benthic zone," and the residents are collectively referred to as "the benthos."

When my students pull nets through littoral vegetation on the sheltered shoreline of Malletts Bay, they find far more species than we dredge up from one hundred feet deep in Lake Champlain. In addition to snails and bloodworms also found deeper in the lake, we pick up dragonflies, damselflies, caddisflies, beetles, and even the occasional stonefly. Higher oxygen content and growing plants support a far more diverse community of invertebrates than can exist at depth.

The rocky, wave-washed shoreline of Lake Champlain's Oak Ledge Park provides a sharp contrast to weedier Malletts Bay locations. Each September, my Community Ecology students visit the park, and picnic tables become our "laboratory." We wade into the shallows and each of us emerges after netting a chunk of rocky rubble along with all of its residents. The wind-driven waves provide enough water movement to support organisms more typical of river habitats. In addition to many species typically found in lakes, we find river invertebrates including water pennies, riffle beetles, net-spinning caddisflies, and huge aggregations of the caddisfly *Neophylax*.

Although we may expect to find only lentic organisms in standing water habitats, no one told the organisms and many refuse to respect our tidy categories. Each fish or insect finds its niche where conditions are right for survival, growth, and reproduction. The organisms in this part of the book are generally found in lakes, but many species are widespread in diverse aquatic habitats. Whole families of organisms are often even more broadly distributed with representative species occupying several habitats. So this part of the book's ensemble is found in lakes, but don't be at all surprised if you also find some in your favorite river.

WATER SCORPIONS—
UNDERWATER ASSASSINS

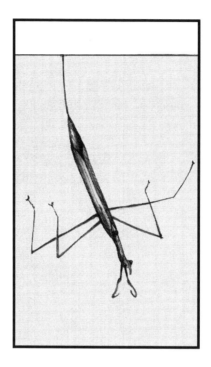

Recently, my daughter participated in Odyssey of the Mind, a creative problem-solving competition devoted to ingenuity and teamwork. As an entomologist, I was thrilled to learn that the program calls its highest award the *Ranatra fusca*. Not only was the award named for an insect, but an aquatic insect, and a particularly fascinating one to boot.

Ranatra fusca is the Latin name for a water scorpion, a creature little known to the general public but very familiar to those of us who wield nets in ponds. This insect bears only a passing resemblance to real scorpions (which are arachnids, not insects). It does sport what looks like a prominent tail (more about that later) but lacks any sort of stinger. It can, however, be quite lethal—if you happen to be an aquatic insect, tadpole, or even a small fish.

Humans have nothing to fear from water scorpions. Unlike their larger cousins, the giant water bugs known as "toe biters," water scorpions are not known for biting on toes or other human parts. Besides, they live in tall weeds where entanglement would be a far greater health hazard than insect nibbles.

Despite measuring in at more than three inches, these amazing predators are easily overlooked by casual observers, and by their prey. They combine stealth with uncanny camouflage. They hold their long, stick-like, brown bodies parallel to the vertical stalks of plants and can remain perfectly still while breathing through a snorkel.

Unlike the snorkel I use when observing aquatic insects, water scorpion snorkels sprout from their rear ends. This is what looks like a tail and inspired their common name. Rather than a single tube, a water scorpion snorkel consists of two half cylinders held tightly together to create an airtight pathway (at the insect's death, these often separate). The snorkel conveys air to spiracles, or breathing holes, on the abdomen.

The snorkel is impressive enough, but the water scorpion has another breathing trick in its repertoire. Once air reaches the spiracles, water-repellent hairs trap it within a bubble. What this means is that the water scorpion has an on-demand scuba tank. When fully submerged below the water's surface, it can still breathe from this reserve of air.

But wait, there's more. When winter ice blocks access to the surface, the bubble switches function from a scuba tank to a gill. Oxygen diffuses continually from the water into the bubble. The water scorpion can survive on this diminished air supply, aided by a dramatic reduction in metabolic rate as the temperature drops.

Thanks to these adaptations, water scorpions can wait for long periods until their next meal swims by. Then they give a nightmare performance of the old AT&T slogan "Reach out and touch someone." Long raptorial front legs whip out like jackknives and firmly snatch the hapless prey right out of the water column. Things just get more fiendish from there.

Actual scorpions use their claws to quickly tear up their prey and thrust the fragments between their jaws. As painful as that may sound, death by water scorpion is a worse way to go—a drawn-out and gruesome affair. The insects skewer their prey with a pointed mouthpart and suck out their fluids as if through some sort of barbaric drinking straw. At the end of the meal, all that's left is an empty husk.

When my colleague Scott Lewins takes his Saint Michael's College students out for their first day of insect collecting to Gil Brook in Winooski, water scorpions are high on the list of coveted specimens. For the budding entomologist, what could be cooler than a large insect that looks like a stick, preserves very well, and is easy to identify? We have specimens in our teaching collection that date to the 1990s with their spindly limbs and separated breathing tubes still intact.

Back on land, it was gratifying to see the excitement on my daughter's face and on the faces of her teammates when, after three years of competition, they received the *Ranatra fusca* award. An insect with Swiss-army-knife appendages, scuba gear, and camouflage is the embodiment of out-of-the-box thinking.

LANDSCAPE ENGINEERS

When my sisters visit from Ireland, I try to play tour guide, but I'm occasionally at a loss for what to do next. During a visit in the late 1990s, my sister Grace said she would love to see a beaver. At that time, I lived close to a beaver pond and often quietly waited for beaver sightings. Alas, the rodents failed to cooperate for Grace's visit, although she was able to see their engineering work. I was disappointed for her, but not surprised. Many of my own encounters ended with at most a fleeting glimpse, and a loud slap of a leathery tail on water.

When I returned to the pond years later, the beavers had departed—but the dam remained. Seven feet tall and made of sticks and mud, the dam had an upstream arch that spanned more than fifty feet of stream valley. According to Tom Wessels in his book *Reading the Forested Landscape*, old beaver dams can last for decades. Wessels points out that beavers engineer more than mere dams, however.

"Beavers are the only animals, other than humans, who will create entirely new ecosystems for their own use," he writes. "And often, like humans, once they have depleted an area's resources, they will abandon their holdings and move on."

Beyond dams and lodges, beavers sometimes dig canals to aid their movement, as well as to float saplings and limbs to stock their underwater larders. Some tree species die after being submerged in beaver-made ponds, becoming habitat for woodpeckers and other wildlife. Some favored food trees, such as big-toothed aspen, resprout from their stumps, producing early successional habitat and multiple delectable stems for beavers to eat.

Eventually, when the beavers exhaust their supply of food trees within easy distance from their pond, they seek out new wetlands. In their wake, they leave an enriched ecosystem that benefits other wildlife.

In areas where beavers can resettle along the same water system, their ponds can serve as aquatic habitats for decades. Well-established beaver populations provide a complex combination of active ponds, abandoned ponds, and beaver meadows in various phases of succession. These create a diverse set of habitats that increases biological diversity across the landscape.

Abandoned beaver ponds accumulate silt and fallen leaves, forming rich soil that eventually fills the pond basin. Light from the canopy gap and well-watered, rich soils support lush communities of grasses and wildflowers called "beaver meadows," which store an abundance of carbon. This soil continues to build as grass grows, lives, and dies. Beaver meadows may remain open for decades, even if the beavers don't re-flood the area, due in part to a lack of mycorrhizae necessary for tree colonization.

Another important physical impact on the landscape is the animals' effect on groundwater. Beaver ponds are far deeper than undammed streams, and pond water saturates surrounding soils. This raises the groundwater table for some distance around the pond. The pond, together with the higher water table, stores a huge volume of water. During dry spells, water seeps from the pond and riparian water table to sustain streamflow. Rainstorms that might otherwise have quickly scoured and eroded stream banks recharge the pond and water table.

Flooding from small storms is contained by the combined water storage capacity, and erosion caused by larger storms is reduced.

I was surprised to learn that beavers also live in—and engineer—salt marshes. In his studies of beavers in Washington State, researcher Gregory Hood found the animals constructing dams in tidal marshes that were submerged completely during high tide but retained water as the tide went out. These dammed marshes provided far more habitat for juvenile fish than similar marshes lacking beaver dams. Beavers sometimes pay a high price for their marine existence, however. In a 2019 article, Ben Goldfarb (author of the popular book about beavers, *Eager*) described beavers suffering and dying from salt intoxication after consuming too much seawater.

Although my sister and I didn't see beavers during our rambles in Vermont, Grace may now have some hope for beaver sightings closer to home. In Scotland, the reintroduction of beavers in 2009 has resulted in an increase in lake levels, higher retention of organic matter in streams, and reduced flooding. In 2016, the Scottish government deemed the reintroduction a success, and in 2019 declared beavers a protected species. It seems Scottish beavers are having positive impacts, much like their North American cousins.

PHANTOM MIDGES—
LATE-NIGHT FEEDERS

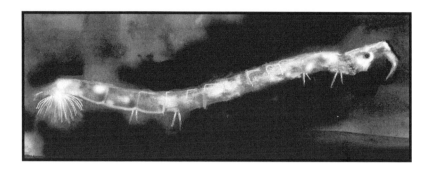

Phantom midges are among the most common, but least seen, planktonic insect larvae in lakes and ponds. These members of the genus *Chaoborus* earn the "phantom" moniker from both their unique appearance and their unusual behaviors.

Measuring nearly an inch long, phantom midges are virtually impossible to see. Their almost transparent bodies warrant another common name: glass worms. Other than a small eyespot, their most visible structures are two pairs of air-filled sacs. Those sacs facilitate the midges' ability to move through the water vertically.

Phantom midges rise to the lake surface for midnight snacks. Their late-night tendencies are another reason humans rarely see them. Finding them, however, is reasonably easy, particularly if you are camping or living near a lake or pond. To join the hunt, you need a bucket, a headlamp, and a pair of nylon stockings.

Head to your nearest lake well after dark and dip some water into your bucket. Pour the water through the nylon stockings—and repeat. The more water you pour through the stockings, the more plankton you will accumulate. After you've collected the plankton, add a little water to your bucket, and turn the nylons inside out to rinse the concentrated plankton into your bucket. A white bucket helps the plankton stand out.

Why is nighttime the right time for phantom midges? It comes down to food and predators. Phantom midges occupy the middle of the food chain. At the food-chain base, algae harness solar energy to string carbon atoms into sugars and starches. Algae are eaten by small zooplankton, including water fleas and mosquito larvae. These tiny animals are eaten by phantom midges, which in turn are eaten by small fish.

Because the sun drives the system, most food is near the surface, where light penetrates—and phantom midges must be up there to feed. But the light presents a problem: fish that eat phantom midges are visual predators, so visible midges become dead midges. The solution for phantom midges is to dine near the surface at night, and spend daylight hours at the murky lake bottom, where hiding is easier and low oxygen concentrations exclude most fish. Phantom midge species that don't migrate are found only in fishless lakes.

Reasons for phantom midge migration are well known, but how this feat is accomplished remained mysterious until recently. In January 2022, Evan McKenzie from the University of British Columbia and colleagues described a unique buoyancy control mechanism in these larvae that comes down to acidity. After a night's foraging near the surface, the larvae increase the acidity surrounding their air sacs, which causes bands of a protein called resilin to contract, compressing the gas in the sacs. Compressing the air sacs makes the organism denser than water, and the larvae sink. The system reverses to alkaline at night and—presto!—the larvae bob back to the surface.

Phantom midges are superbly adapted both for vertical migration and for capturing prey. In deep lakes, they travel hundreds of feet during a daily commute that takes them from the security of the lake floor to open water densely stocked with delectable snacks.

When the larvae arrive closer to the surface, they hunt until dawn drives them back to depth. Like other true fly larvae, phantom midges lack legs to tackle prey. Instead, their antennae have evolved raptorial function that would make Hugh Jackman's Wolverine claws in *X-Men* look positively tame. Part grappling hooks, part impaling devices, these multi-tool antennae work with the larva's mandibles to pierce and crush prey items that will nourish larvae into their next life stage.

When the time comes, the midges pupate until it is time to emerge as adult midges. These adult midges are nonbiting but may indulge in some nectar sipping. It is for this brief portion of their life cycle that they may finally come to the notice of people—as clouds of small flies on a summer's evening. These flies live for a few days, then lay rafts of eggs on the water surface to complete their life's mission before succumbing and, perhaps, at last becoming fish food.

MORE MUCK—THE JOYS OF LAKE FLOOR SAMPLING

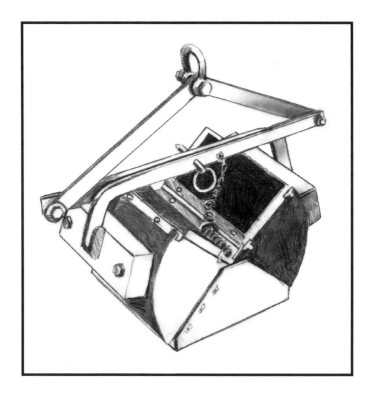

Once final exams are over and we wrap the academic year, my friends sometimes assume that I then retire to a deck chair, consume bonbons, and work on my tan. In reality, most academics use summer for various scholarly pursuits. While some work poetic magic with words or pore over ancient manuscripts, I trade my cap and gown for muck boots, a personal flotation device, and lots of sunscreen!

In June 2019, my students and I joined Patricia Manley and Tom Manley from Middlebury College on the *Research Vessel David Folger* to sample Lake Champlain's floor. We started our collaboration some years back in Missisquoi Bay, continued on to Saint Albans Bay, and 2019 brought us to the Inland Sea extending between the two bays and separated from the main basin of Lake Champlain by a chain of islands.

Tom uses side-scanning sonar to refine bathymetry maps, accurate lake floor renderings that correct and dramatically improve on earlier surveys. Patricia's focus is on sediments, which is also why my students and I are along. Three teams of scientists answering different questions from the same research vessel appeals to my sense of efficiency, and I love being on the lake.

Patricia's sediment sampling technique is exactly how I collect macroinvertebrates, the "large" invertebrates from lake beds. We drop a heavy steel Ponar dredge overboard, and its momentum digs it into the silty lake bottom. The dredge's open jaws frame a 9" square. When it hits bottom, a spring-loaded pin releases allowing a yank on the cable to close the jaws and take a bite out of Lake Champlain's muddy floor. A winch winds up a hundred or more feet of cable and Captain Rich Furbush motors on to the next sampling point. I consider this luxury sampling; when I take my turn sampling shallow waters from a small boat closer to shore, there is no winch, and my muscles ache by day's end.

It's a little like fishing with an incredibly unfair advantage: we always catch what we want. Well, we nearly always catch what we want; sometimes we hit a rocky ledge while other times a pebble or zebra mussel props the metal jaws open and our sample washes back into the lake. Most times though the dredge delivers a cube of glistening muck. Patricia's students use syringes and spoons to get all they need to measure sediment characteristics.

The rest of the fabulously gooey layer cake is mine. I encourage my students to "feel the muck"; stick a finger in there and experience the refrigerated lake floor temperature even when the surface is sufficiently balmy for a swim. My interest is in the macroinvertebrates, but what do we mean by "macro," or "large"? The answer is quite simple: we run the sample through a 0.6 mm sieve; whatever gets caught is "macro" and whatever washes through is "micro." It's one of those arbitrary but agreed-upon scientific standards.

A surprising array of organisms can be found, even in the utterly dark "aphotic" zone of Lake Champlain. No light means no photosynthesis and so the base of the deep-lake floor food web consists of whatever living or dead material settles down from above. Unlike the flowing water that keeps many rivers perpetually murky, gravity in the standing water of lakes and ponds drags even very fine particles down enriching the lake floor.

A good portion of the deep community makes a living by filtering fine particles of organic material from the water. Burrowing mayflies, lamp mussels, fingernail and pea clams, and more recently arrived European zebra mussels pump Lake Champlain through their bodies and dine on whatever they filter out. The lake water is cleaned as a result, and my students have certainly seen the effects on water clarity of even a single mussel in an aquarium. Captive mussels don't last long, however; they might clear a fish tank of material that clouds the water in a day or two, but with all the food consumed they die, stink, and as they decompose they kill the other residents of your tank. Besides, our native mussel populations are at risk and should not be brought home; zebra mussels are invasive and should not be moved.

Our samples also include a healthy community of snails which is once again dominated by Europeans: two species in the genus *Valvata*, or "valve snails," and another invasive called *Bithynia tentaculata* or the "faucet snail." These snails are in some ways analogous to a pair of Vermont's state symbols: red clover and honeybees; Europeans that have been here so long that many consider them naturalized; not unlike myself I suppose.

Exploring Lake Champlain's floor in this way reveals the underpinnings of the food web that extends from humble insects, mollusks, and crayfish to perch and lake trout right through to ospreys, bald eagles, and ourselves. Identification takes weeks, providing valuable experience for budding field biologists. And by the time I have had very nearly enough of burrowing mayflies, bloodworms, and muck, it'll be time to teach fall courses. They say a change is as good as a rest; I couldn't agree more.

A LAKE TSUNAMI

The sharpest contrast between rivers and lakes is in water movement. While rivers flow inexorably downhill, lake water movement is more subtle. But anyone who weathered a storm on a lake can attest that less consistent water movement does not mean zero water movement. Seasonal differences in water movement record history in lake floors that one professor has used to document a Lake Champlain tsunami.

Relatively still lake waters settle out the sediment carried downstream by turbulent rivers. Sand and silt accumulate at rates determined by precipitation and river flow patterns across the lake's entire watershed. Spring snowmelt swells rivers dramatically washing impressive quantities of material from the landscape. Because of this, proportionally more sand and silt get deposited early in the year. Wind and wave action keep the very finest of materials suspended in lake water through fall. Lake ice puts a lid on things in midwinter, allowing very fine clay-size particles to settle out.

Seasonal differences accumulate on lake floors like tree rings and can be seen as paler and darker stripes called "varves" in lake sediment cores that geologists use to study past lake conditions. Cores are long columns of lake-floor mud sampled by driving a hollow pipe deep into the lake bed, sometimes from a boat, other times through thick winter ice.

In Lake Champlain, cores have revealed a pro-glacial freshwater lake called "Lake Vermont" that was dammed to the north by retreating glaciers and drained south through the Hudson Valley toward what is now New York City. Once the glaciers melted back past the Gulf of Saint Lawrence, seawater filled the depression in the landscape caused by the weight of the recently departed ice forming the "Champlain Sea." Lake Champlain formed once the landscape rebounded from crushing ice weight and emerged from below sea level.

Sea caves, sand dunes, and even an isolated subspecies of marine beach grass (*Ammophila breviligulata ssp. champlainensis*) remain from the days of the Champlain Sea. In 1849 a beluga whale skeleton was uncovered in Charlotte, Vermont, by railway workers. After a successful campaign by the children of Charlotte, the whale was declared the official state fossil.

The mud samples we take using metal dredges barely scratch the surface of this deep history and lack the penetrating power of sediment cores. And even cores yield a limited perspective because they sample just one point on the lake floor at a time. To get the bigger picture, Dr. Patricia Manley uses some powerful technology hauled behind Middlebury College's research vessel.

Patricia Manley and her longtime collaborator and husband Tom Manley broadcast "compressed high intensity radar pulses" (CHIRPs) into the lake water and the sediment below. Radar pulses bouncing from subsurface varves are recorded by detectors and used to build three-dimensional X-ray-like images of sediments laid down over centuries. These images take us back to the end of the most recent ice age and suggest that tsunamis are part of Lake Champlain's history.

Varves are not just monotonous, layers upon identical layers of crud. Just like tree rings, there's quite a bit of variability from year to year. Melt from one year's deeper snowpack may leave a thicker sandy layer; a year with little lake ice may obscure the clay layer; and extended ice and reduced melt, as in 1816, the infamous "year without a summer" would leave a unique signature. All of these basin-wide patterns conspire to produce a barcode-like pattern of consistently varying relative varve thickness across the lake floor.

To the trained eye, departures from this barcode stand out. Patricia Manley and her collaborators noticed a particularly spectacular departure from typical just south of the Bouquet River on the New York side of Lake Champlain. Everything looked normal for the most recent nine hundred years or so. But a dramatic upheaval was evident approximately a thousand years back. A large swath of sediment was missing from a shallow slope. Above and below this gap in the record, there was a normal accumulation of older sediments just sitting there and behaving as sediments do.

The mysterious lack of sediment was solved by looking deeper in Lake Champlain. Tons of sediment went downhill in a catastrophic underwater mudslide that upended many decades of accumulated sediment causing the historic barcode to read backward from older sediment down through younger varves that had been rolled under. Above the topsy-turvy misplaced sediment pile, additional centuries of sediment had accumulated in the normal way, encapsulating the record of the mudslide.

So much mud had cascaded downhill that Dr. Manley calculated the resulting pressure wave would have produced a tsunami. But how large is a tsunami? According to Dr. Manley, the wave would have been sufficiently large to wash over Shelburne Point and into Shelburne Bay.

What might cause sediment that had sat quietly for a millennium to suddenly and destructively take flight? According to Dr. Manley, an earthquake was the likely culprit. In the process of researching her findings, she found published records of above-water landslides in the Western Quebec Seismic Zone. The timing of these landslides matched up with that of the Lake Champlain mudslide.

And while a tsunami risk is not going to keep me off Lake Champlain, I find it fascinating that forces beyond our current perception have shaped our lakes in ways I never would have guessed.

GIANT WATER BUGS—
BEST FATHERS IN THE POND!

While sitting poolside with my children, another parent hustled her son out of the water because of a swimming cockroach; I was intrigued. The "cockroach" turned out to be a giant water bug (family *Belostomatidae*), the largest of the hemipterans or true bugs. And that mother's instinct was a very good one.

Another common name for these spectacular insects is "toe biter," and more than a few swimmers know why. When a giant water bug minding its own business on a pond floor gets stepped on, it uses what defenses it has and bites the offending toe to convince its owner to move on.

And the offending swimmer would tend to move on quite rapidly. While I hate to demonize any insect, particularly one that rarely bites people, giant water bugs deserve respect. Their bites have been described as intensely painful or even excruciating and sometimes lead to numbness of an entire limb. Symptoms vary but may persist for up to five hours, typically without lasting damage.

You may wonder why a large and robust insect needs such potent weaponry; the explanation lies in how they feed. Like all true bugs (Hemiptera), giant water bugs pierce their food sources with a penetrating beak that injects liquids and later sucks back a fluid meal. Giant water bug venom includes cell-destroying toxins and enzymes to break the victim's proteins down to a conveniently drinkable soup.

Giant water bugs hang upside down in aquatic vegetation, their pointed abdomens close to the water surface. When the need arises, they poke snorkel-like breathing tubes up for air with scarcely a motion. They patiently wait until an errant fish, or perhaps tadpole, swims too close, at which time the bug goes from zero to sixty in a flash.

They use incredible speed and swimming skill to pounce on prey, and while grabbling with all six legs, they engage in a brief struggle that is sometimes reminiscent of a rodeo ride. The struggle is brief, because a single puncture wound delivered by a pointed proboscis is all that is required to subdue prey. It is less the physical wound and more the venom that delivers the *coup de grâce*. Giant water bugs dine on insects, amphibians, turtles, and snakes, and there are even reports of them taking the occasional duckling.

I have read giant water bugs described as having "powerful pincers." This is an understandable confusion, because their very strong forelegs exert a tight grip. And let's face it, when an insect of this size grabs your child's toe, it's not the time to point out the lack of pincers.

Giant water bugs also have a gentler side and are known to be among the best parents, and specifically fathers, of the insect world. Depending on the species, two approaches to fatherhood are common. Females of the larger species (subfamily *Lethocerinae*) lay eggs on aquatic plant stalks right above the waterline, where they are tended by the males. The fathers defend the eggs against predators, shade them with their bodies, and make excursions to bring water to moisten the eggs.

The smaller species (subfamily *Belostomatinae*) practice what amounts to male pregnancy and even play rock-a-bye, baby with their developing offspring. Mother water bugs lay eggs directly onto the backs of their mates. In some cases where male back space is in short supply, females will even try to get between a happy couple to lay eggs on any father's back; they might be considered the cuckoos or cowbirds of the insect world.

Regardless of egg source, "pregnant" males alter their typical routines to ensure successful egg development. Instead of hanging in the weeds awaiting prey, brooding males make risky trips to the surface and push their backs above water to improve air supply for their precious cargo. While submerged, they use a combination of pushups and a rocking motion of the entire body to flush fresh water past the eggs, thus removing wastes or sediments that might impact survival.

And their efforts are generally successful. Eggs tended by males above water, or attached directly to the backs of males, have dramatically higher hatching rates than neglected eggs.

In one study, Noritaka Ichikawa of the Himeji City Aquarium in Japan determined that fully 94 percent of eggs on plants attended by males successfully hatched. All of the neglected eggs in the study simply dried up and died.

So critical is the father's role that some females will resort to nefarious means to ensure their eggs are cared for. When males are small in number, marauding females will sometimes destroy eggs from other females being tended by males, and then mate with the newly available male. The males do what they can to defend their first nest, but if unsuccessful, then I guess a new nest is better than no nest.

Should you be concerned about protecting you and yours from apparently ferocious pond and river dwellers? I searched to see who got bitten. Of the seven victims discussed in case studies from Brazil, five were professional biologists bitten while working; the at-risk demographic it appears, is me. So unless you are planning on joining my strange profession (and you would be most welcome), you may well be fine. After a quarter century in without a bite, I'll take my chances.

UPSIDE-DOWN AQUATICS

I had just finished my safety talk to some middle school students when I heard a bloodcurdling scream. In many years handling aquatic insects and other small water creatures, I have never been wounded. Crayfish have once or twice gotten hold of me but never drawn blood. So I was quite surprised to hear through the minor chaos that a student had actually been bitten.

There were no crayfish where we sampled in Winooski floodplain ponds and only one likely candidate to produce such a scream. It was the reason I had specifically warned my students to use forceps.

Backswimmers (*Notonectidae*) are also called "water bees," and according to Dr. James Diaz of the Louisiana State University School of Medicine, they frequently deliver a painful bite when threatened. Unlike their shorter-beaked cousins the water boatmen (*Corixidae*), backswimmers have long, segmented beaks that easily penetrate human skin. They use these fearsome beaks to dispatch prey that range in size from insects smaller than themselves all the way up to small fish and amphibians. The chemicals that paralyze their prey likely cause more pain to occasional human victims than does the physical injury.

As their common name indicates, backswimmers swim on their backs. This is not an occasional convenience as might suit you or me; this is the only way these insects swim. Everything about their anatomy, coloration, and behavior fits with this unusual behavior.

Many aquatic cousins of backswimmers, including water striders, water boatmen, and more distant relatives such as diving beetles, use countershading to be less visible. Their backs are dark to match the pond floor when viewed from above; their undersides are pale to blend with the sky. Backswimmers buck this trend because their dark "undersides" are viewed from above. Their paler pearlescent backs obscure them from any potential predator coming from below.

In common with other insects, backswimmers' six legs grow from the "underside" of the thorax, except of course that they face up toward the water surface. When they feel the need to dive, a pair of powerful oar-like back legs propel them to the pond floor where they can hang on and remain submerged for extended periods.

Staying submerged is a choice in spring, summer, and into the cooler months of autumn, but what happens in winter? Thick ice entombs backswimmers along with many of their aquatic neighbors. One might expect these insects to hibernate, but nothing could be further from the truth. While we skate across a frozen lake, or drill a hole to fish, backswimmers remain active below, hunting prey and growing.

So how do they survive in such a cold, poorly oxygenated environment? Partly through economy. In colder temperatures, their metabolism and corresponding need of oxygen is somewhat diminished. They also carry bubbles with them—these are attached to grooves in the abdomen that are surrounded by hydrophobic hairs. The bubbles are replenished by oxygen diffusing from the water.

The bubbles also serve as a means of passive locomotion. Backswimmers are members of one of only three insect families that use hemoglobin to store large quantities of oxygen. They can shunt oxygen from their hemoglobin into their external bubbles and back again: when they shrink their bubbles, they sink down in the water; when they emit oxygen from abdominal tissues back into their bubbles, they rise upward.

This remarkable trick allows them to maintain a resting state of neutral buoyancy. It reduces their energy costs and thereby helps them survive harsh winter conditions.

And speaking of enduring harsh conditions, what about that middle school insect bite victim? He jerked his hand away, and the offending backswimmer landed at the feet of a colleague. His hand swelled up a little, but he was a trouper and even seemed to enjoy the attention as we cleaned and applied antiseptic and a Band-Aid to his hand. He jumped right back into the lesson with enthusiasm and paid a lot more heed to my warning to use forceps.

THE UNDER-ICE FOOD WEB

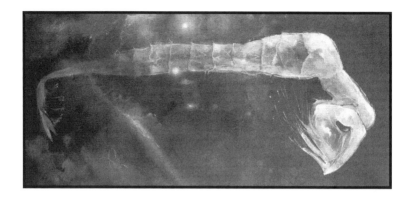

Earlier this winter, I took to the pond ice—not to skate, but to peek below the surface.

Although lake ecologists once considered the plankton in frozen lakes to be dormant during winter, recent studies reveal that the plantlike, microscopic phytoplankton (which move with the lake's currents) and animallike zooplankton remain active below the icy surface.

In data collected from more than one hundred lakes, Washington State University Professor Stephanie Hampton found that while the base of the under-ice food web is reduced compared to summer, it is certainly not dormant. Hampton's research suggests that the phytoplankton supporting lake food webs is reduced by 80 percent compared to summer, while one level up the food web, the abundance of zooplankton falls by 75 percent in winter lakes.

Still, there is enough activity within that web to support a variety of winter lake life, something New England's optimistic ice-fishing community has long understood. The fish, however, are certainly not hanging around in vain hopes of a sudden shaft of light through the ice and meals provided by humans like manna from heaven. They rely on a natural food web supported by phytoplankton, which harnesses the sun's energy to make food. These phytoplankton are consumed by zooplankton, which are, in turn, large enough to provide tasty morsels for fish.

Just as the fish are not swimming around purely for the benefit of winter anglers, the zooplankton are not in it for the sake of fish. While fish and other organisms evolved as predators, there is no Darwinian imperative to become prey. Zooplankton evolved to avoid being eaten, and behavioral adaptations keep many species out of harm's way.

One of these adaptations is akin to "the largest daily migration on Earth," according to Ariana Chiapella, a research associate at the University of Vermont's Rubenstein Ecosystem Science Laboratory.

During the summer, many zooplankton species spend their days out of sight on lake floors, even though most of their food is near the bright surface. At night, protected from fishy nemeses by the veil of darkness, these zooplankton pop up to feast. The phantom midge, one of those nightly migrants, makes a living hunting smaller zooplankton by night and settled into the mud by day.

Most of the research documenting these nocturnal migrations has been done during summers. Chiapella and collaborators in Jason Stockwell's UVM research group wondered if this and other daily migration also happens under winter's ice, when instead of ponds being warmest at the surface, as they are in summer, the opposite is true. And so, one March morning, they took to Shelburne Pond with ice augers and plankton nets.

They sampled around the clock from just beneath the ice to more than fourteen feet below the surface. To obtain their samples, the researchers lowered a pipe through ice holes and pumped five gallons of pond water up into buckets from each of several depths. After first sampling small volumes for microscopic phytoplankton, they poured the remaining water through fine-meshed nets to collect the larger zooplankton.

In their first midday samples, there was no sign of phantom midges at any depth. What was common in the midday samples, however, was *Daphnia mendotae*, a water flea species commonly eaten by phantom midges. Nighttime samples told a different story. The phantom midges came up from depth at twilight and were found in the deepest samples. By midnight they were abundant at all depths from fourteen feet deep to just beneath the ice surface. And their prey, the water fleas? They went deeper at midnight and came up closer to the ice after sunrise when the phantom midges had gone to ground.

I asked Stockwell why the phantom midges did not simply track the water fleas and stay up both day and night. He reminded me that fish rely on daylight to find food, including phantom midges, so the midges dive because, as Stockwell said, "It's better to be hungry than dead."

These studies confirm that the base of the lake food web is alive, well, and migrating year-round, but perhaps on a calorie-restricted winter diet. My diet tends to follow an opposite pattern!

THE WORLD TURNED
UPSIDE DOWN

As I waded in Lake Champlain at North Beach one summer, a fellow bather explained that just a little farther out, refreshing spring water would cool my feet. I have heard the same old wives' tale repeated at Lake Arrowhead in the Pennsylvania Poconos and in Sandy Bay on Lough Ree in the Irish midlands. The explanation of colder, deeper water is not quite as simple as coincidentally occurring springs "just a little farther out." And as the seasons change, the same explanation turns the lake world upside down.

Textbook diagrams of summer lakes show layering of warm water on top, cold water below, and a thermocline, or transition zone of rapid cooling water between the two, that many scuba divers can confirm. During my scuba certification dive into a water-filled Ohio quarry, the instructor took us down about twenty feet to experience the thermocline firsthand.

After submerging we followed an anchor line until we felt cold water up to our hips, while our torsos remained in distinctly warmer water above. One by one, we joined the instructor for a procedure that may sound like hazing but is actually an important test of preparedness. We were brought deeper and told to remove and then replace our face masks.

Classroom training had warned that a rush of cold water to the face causes a gasp reflex that can deliver a lethal lungful of water to those poorly prepared. And the deep water of an Ohio quarry was certainly cold enough to cause sharp inhalations through my air regulator.

If diving is not your thing, you can demonstrate lake layering using food coloring and glass jars. Fill one jar with cold water and add food coloring; I like blue for cold. Fill the second jar from your hot-water tap and add red food coloring. Now take an index card or envelope and cover the warm jar. Flip the warm jar upside down on top of the cold-water jar and slide the card out. The cooler denser water will remain in the bottom jar and the warmer, less dense water will sit politely on top for quite some time.

As seasons change, sun-warmed water that spent summer on top of a lake or pond cools gradually, and just as gradually gets denser. At some point the water on top is as dense as, or denser than, the water below. Once that point is reached, the surface water sinks and mixes with the underlying water in a process called "lake turnover."

Sometimes the cold winds of a winter storm deliver the final push to turn a stratified summer/fall lake into a well-mixed winter lake. An ice pack added to your baby food jar can stand in for a cold winter storm for this demo. My time-lapse YouTube version of this has been viewed by hundreds over a period of several years—it's not exactly a viral video.

My friend Patrick Standen reported from an October 6 Lake Champlain swim: "air temperature 58 degrees; water temperature 60 degrees." I guessed that the lake had not yet turned over. The University of Vermont's research vessel captain, Steve Cluett, shared lake temperatures from the same time frame. On October 1, the temperature from the surface down to about 50 feet cooled from 62 to 61 degrees Fahrenheit and as Steve's thermometer sank to 80 feet it cooled quickly to 54 degrees. From there the water cooled slowly measuring 51 degrees at 125 feet depth.

An 11-degree difference between surface and 125 feet deep was enough to confirm that the lake was still in layer-cake mode: warm surface, cooler deep, with a rapid transition in between. And this status quo seemed to continue for the next several days. But a closer look at the numbers revealed changes afoot: the temperature difference between shallow and deep was shrinking.

By October 10, the difference was 7 degrees and just 5 degrees on October 20. By October 22, the surface was just 2 degrees warmer than the deepest reading and the zone of rapid temperature change had disappeared; the lake had turned over. The surface water by then was about 54 degrees Fahrenheit; cold enough to give even my hardy friend Patrick pause.

Lake turnover is more than just a curiosity for scuba divers. Nutrients are brought to the surface and oxygen is brought to depth by the rapid transition. According to Mary Till's 2014 report on lake trout and climate change in the Adirondacks, lake trout depend on turnover to deliver oxygen to their cooler, deep-water haunts where oxygen can otherwise reach critically low levels.

And so, next time you find your feet pleasantly cooled during a summer swim, remember that the cold below contributes also to healthy fish populations. And you can determine the wisdom of explaining thermoclines and lake turnover to believers in cold springs "just a little farther out."

PART V

WHAT EMERGES FROM THE DEEP

The mayfly lives only one day. And sometimes it rains.
—GEORGE CARLIN

Mayflies, "Ephemeroptera," their name says it all: we use them as our favorite analogy for the brief or ephemeral. But the analogy sells the mayfly short. The brief adult existence of a mayfly would not be possible without a year or more of feeding and preparation beneath the waters of a river, lake, or pond. And all of this feeding means that an emerging hatch of mayflies, sufficient to show up on regional Doppler weather radar, represents a huge movement of nutrients and energy from aquatic habitats to dry land.

Feeding and growth as a larva or nymph for a year or more is the case for the majority of aquatic insects that eventually emerge as adults. And even noninsects such as snails, crayfish, isopods, and amphipods with strictly nonemergent lifestyles contribute to the terrestrial food webs, nonetheless.

Just as surely as leaf fall and erosion from the landscape contribute to aquatic food webs, freshwater habitats contribute to terrestrial food webs. When great blue herons, kingfishers, dippers, and bears feed from a river or lake, they move nutrients eventually to dry land.

Water bodies cannot be viewed in isolation. Each river and lake is intimately connected to the surrounding landscape and can even serve as a conduit that brings marine nutrients in the form of salmon and trout upstream to feed terrestrial food webs. This part of the book covers some of the links between freshwater and dryland habitats.

AQUATIC LIFE EMERGING

As I sieved a sample on the banks of Lewis Creek one afternoon, a hard object struck me on the cheek and splashed into my sampling basin. I expected to find a twig or chunk of sycamore bark, instead I was greeted by a fast-swimming giant water bug.

A slap in the cheek was a graphic reminder of the connections between freshwater and terrestrial habitats, not that I felt I needed one. Water, leaves, soil, and woody debris are regularly deposited into water bodies and, in appropriate doses, are essential contributions to river and lake habitats. But what moves in the other direction? What contributions do water bodies make to dryland habitats?

Numerous studies in Alaska have tracked food movement from the ocean, up into streams, and eventually to dry land. Salmon that hatched from Alaskan streams do 90 percent of their growing at sea and as they swim home in large numbers they represent an enormous influx of nutrients. Grizzly bears binge while the salmon runs last and discarded fish scraps vanish rapidly into the waiting mouths of scavengers large and small. And we all know what bears do in the woods: they fertilize the food base of woodland ecosystems, and for a while each season this fertilizer comes directly from the Pacific Ocean by way of rivers and streams.

Fish, whether caught by bears, humans, or other animals, provide valuable nutrition. Some species like ospreys, bald eagles, kingfishers, and herons depend on fish for most of their diet. But because of sheer abundance, other, much smaller organisms also provide large links to terrestrial habitats.

Aquatic insects spend most of their lives underwater. They emerge briefly as adults to disperse, mate, and then deposit their offspring back into the water, frequently in the water body they came from. As you'll see in "Cloudy with a Chance of Flies" and "Burrowing Mayflies," some of those insects occur in truly vast numbers and are essential for the success of swifts, swallows, bats, spiders, and other terrestrial insectivores. In a very real sense, these insects return freshwater nutrients to the terrestrial environment, completing the loop that started with leaves falling into rivers and lakes.

When insects emerge after feeding directly on Alaskan salmon carcasses, or on the bacterial and fungal biofilm that forms as the fish decompose, they are tiny, winged smugglers of Pacific Ocean food to the stream-adjacent woods. But many freshwater invertebrates do not share the emergent lifestyle characteristic of insects. The list of freshwater invertebrates that spend their entire life cycle underwater is long and includes snails, crayfish, flatworms, isopods, and amphipods. But even these organisms, together with many insects find indirect ways to make it ashore.

Crayfish are on the menu for semiaquatic animals like mink, otter, and kingfishers. Ducks, geese, and swans tend to dine on their share of freshwater snails and other benthic invertebrates. In western North

America, Asia, and Europe, small birds called dippers dive, swim, and walk underwater while foraging for macroinvertebrates. But perhaps the most voracious predators of freshwater invertebrates are fish. Fish, plucked from any river or lake by birds and other predators, are very likely to have grown to edible size by consuming insects and other invertebrates.

Ecosystems are complex, and there is strong evidence that the most complex and diverse ecosystems are more stable and better equipped to struggle through disturbances, anthropogenic or otherwise. During my years as a summer camp nature counselor, I assigned plant and animal names to campers standing in a circle, and we used a ball of yarn passed over and back to symbolize the links between species. When I pretended to remove one "species" from the circle, all felt the tugs on their strings.

It is possible to construct at least hypothetical links between every aquatic organism and dryland communities, but this part of the book focuses on some that make their emergence felt. Whether by directly posing an inhalation risk to us (and to themselves), or by having a strictly emotional impact by appearing ferocious or even lovable, the highlighted organisms are just a small sample of the unique and diverse life-forms to be discovered and admired as they emerge from your nearest freshwater haunt.

CLOUDY WITH
A CHANCE OF FLIES

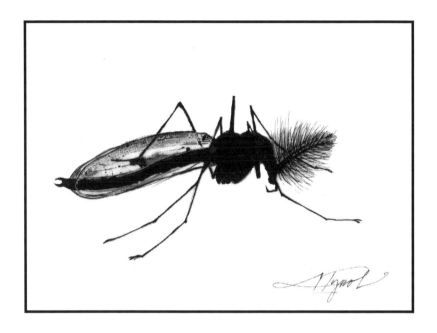

Clouds of tiny insects, rising and falling hypnotically along lakeshores, contribute to the ambiance of warm summer evenings. My recent bike ride was interrupted by a lungful of this ambiance.

If you find yourself in a similar predicament, you might wonder what these miniscule flies were doing before being swallowed, where they came from, whether they bite, and whether we need these interrupters of peaceful lakeside jaunts. We'll get to these questions, but first, let me say that as an ecologist, I find these insects to be among the most fascinating and important freshwater invertebrates.

Nonbiting midges, in the family *Chironomidae*, are most conspicuous when they hover in swarms near water. Not to be confused with the no-see-ums or biting midges (*Ceratopogonidae*) maligned by

Thoreau, they constitute a diverse family found on every continent. Two midge species are the only known insects in Antarctica. One Antarctic species lacks wings entirely and is therefore, I suppose, at little risk of death by human inhalation.

Whether earthbound or in flight, mating is the reason for midge swarms. Newly emerged adults have a short time to find mates, and dense swarms function like singles bars.

My clumsy encounter with a nuptial swarm did not make much of a dent in local midge numbers. They are abundant. They're also hyper-diverse, with more species in this one true fly family than in all the families of stoneflies combined. Entomologists have described more than five thousand species and are regularly called upon to come up with scientific names for newly discovered species. *Dicrotendipes thanatogratus*, a midge named for the Grateful Dead (*thanatos* is Greek for "dead," and *gratus* is "grateful" in Latin), is one example.

With so many species, it's not surprising that nonbiting midges thrive in all sorts of habitats. Saint Michael's College students and I have found them in every stream, pond, and lake we have sampled, and they often exceed 50 percent of the insects we capture. In ponds, lakes, and deep river-silts, bright red midges called bloodworms use skin-bound hemoglobin to scavenge every trace of oxygen from their stagnant habitats. Dr. Richard Jacobsen studied a midge found only on the backs of mayflies; its life cycle synchronizes perfectly with that of its bigger host.

Nonbiting midges are also diverse in their culinary predilections. They eat nearly every conceivable foodstuff; they can be scavengers, herbivores, predators, or parasites. One species, *Metriocnemus knabi*, feeds exclusively on insect parts in pitcher plants in northern bogs. There are abundant midges grazing algae in salt marshes, consuming leafy detritus in tree holes, and they may well be scavenging from the rainwater in the gutters of your house. These insects are essential in the aquatic food webs that support fish populations, and anglers seek to replicate their delicate forms to lure their catch.

Like many insects, midges grow through a series of larval stages. Slim, legless, translucent larvae hatch from eggs laid in gelatinous masses. They grow quickly and become too big for their exoskeletons.

These split, revealing softer exoskeletons that stretch and harden around their now larger bodies. In streams, where I do much of my research, discarded exoskeletons accumulate behind rocks and logs. Sieving exoskeletons from stream froth is a little like panning for gold. Much information can be coaxed from these discarded shells. Cleaner streams host more midge species and certain species are never found in polluted streams.

After four rounds of growing and shedding, nonbiting midges enter a pupal stage. They may appear somewhat static, but each pupa is a hive of cellular activity. Cells migrate and rearrange, forming compound eyes, six legs, and in most species, a single pair of wings. After a few days, the pupae split, and adults emerge. Some midges hatch only at certain times of day and the rhythm of a stream can be followed by sampling drifting pupal skins through a twenty-four-hour cycle.

The clouds of adult flies are a food source that moves from aquatic to terrestrial food webs, sustaining the swallows and bats that keep real pests in check.

Nonbiting midge swarms persist as long as the weather is warm enough for fly muscles to flap fly wings. Some species hatch early in the season, some later. There are species that hatch once per year; others can produce two or more generations in a season.

So when you find yourself in a swarm of insects this summer, try to appreciate the romance. Perhaps the knowledge that fly love is in the air will make up for the occasional fly in your eye. Or in your mouth. Or snuffed right up your nose!

WATER BOATMEN—
FORAGING BENEATH THE ICE

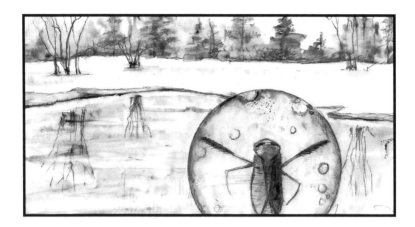

If you get a chance this winter, take a peek through the icy window of a pond surface. You may see water boatmen (order Hemiptera: Family *Corixidae*) clinging to the pond floor. Long oar-like hind legs propel these insects, inspiring their common name. Shorter, scoop-like front legs are used for feeding and singing.

This largest aquatic family of true bugs includes 128 North American species. Unlike their cousins the backswimmers, which swim upside down, water boatmen swim with their winged dorsal surfaces facing up. When not swimming, they cling to submerged objects with their clawed middle legs.

They cling, because otherwise they bob up like corks, and it takes precious energy to swim back down. Buoyancy is a side effect of breathing: beneath their wings and abdomens, water boatmen carry bubbles, which act like tiny scuba tanks.

Fresh bubbles from the water's surface contain 21 percent oxygen, the same as the air you and I breathe. Over time, the insect uses up the oxygen, and the concentration drops. But oxygen from surrounding water continues to diffuse into the bubble, which functions like a gill. Absent ice, the insects surface for fresh bubbles. But in winter, they can remain submerged for extended periods, as colder water holds more oxygen than warm water. Surprisingly, at least one North American species (*Cymatia americana*) can survive encased within solid ice.

Many water boatmen in small ponds and shallow wetlands, however, spread their wings and migrate before ice entombs them. Stephen Srayko from the University of Saskatchewan and colleagues have documented enormous corixid migrations from wetlands to larger rivers. After migration, researchers found more than two hundred water boatmen per square foot in slow-water locations. The scientists estimated that throughout the Prairie Pothole Region, spanning three Canadian provinces and five US states, thirteen thousand tons of water boatman biomass moves from wetlands to rivers. Srayko and his collaborators found that most fish sampled in these rivers dined on corixids, which accounted for up to 97 percent of food consumed.

After overwintering in rivers, water boatmen returned to wetlands to feed and reproduce. Most water boatmen consume aquatic vegetation, but as many as a quarter of the North American species prey on invertebrates, including other water boatmen.

Some corixids generate sound announcing their presence to potential mates and rivals by rubbing file-like foreleg structures against a ridge called a "plectrum" on their cheeks. The sounds resonate through their air bubbles, and the size of the insect and its corresponding breathing bubble affects pitch; bigger bubbles produce deeper songs. The sound of one European corixid in particular has attracted the attention of the folks at the Guinness World Records. The species measures one-tenth of an inch long and generates a 99-decibel sound that's as loud as a passing freight train. Curiously, this species generates sound by rubbing a ridge on its penis against its abdomen, earning it the record for the loudest penis on the planet; nature is stranger than fiction.

I was curious if corixid migrations to rivers happen in Vermont, so in January I grabbed a net and headed to the Winooski River. If vast swarms of water boatmen exist in the Winooski, they certainly eluded me. Despite several net sweeps from the bank, not a single water boatman revealed itself. I contacted Aaron Moore, who monitors river macroinvertebrates each fall for the Vermont Department of Environmental Conservation. He told me that they occasionally get water boatmen in samples, but not in numbers to suggest mass migration.

I had more luck locating corixids in a nearby wetland. After breaking through the ice, my first net dip captured a water boatman. With or without a net, the best way to see water boatmen in winter is to approach a pond or wetland edge and look for movement as these insects go about the business of foraging beneath the ice. I find that tapping the ice surface encourages a bit of swimming.

Whether you knock on the icy door or not, I hope that you are lucky enough to see water boatmen in a pond near you. If you do, give a listen; they might even sing for you.

DEALING WITH DEERFLIES

My students and I were conducting research in the Winooski River floodplain at Saint Michael's College one summer when the buzzing became particularly intense. A brisk walk is enough to outdistance mosquitoes, but deerflies combine fighter jet speed with helicopter maneuverability. And a slap that might incapacitate a mosquito seems to have little effect on these relentless pests.

Deerfly season 2017 started slowly, but by late July there were enough to carry off small children. On trails between wetlands and farm fields, we were dive-bombed by countless, persistent, little winged vampires. Insect repellent did little to repel them. We slapped, feinted, grabbed at thin air, and usually came up empty. It was like *Caddyshack*, but with flies rather than gophers.

The horsefly family *Tabanidae* includes deerflies, along with larger Alaskan "mooseflies," and the greenheads that ruin many a trip to New

England's beaches. Iridescent green eyes that make up most of the fly's head give them their common name. Far more impressive is their bite: they truly hurt. Because greenheads emerge only from salt marshes, by calculating the distance to the nearest marsh, researchers have learned that they travel up to two miles in search of blood.

Deerflies and their relatives risk getting hand-slapped and tail-flicked because humans and other mammals offer the high-protein food source they need to develop their eggs. The gamble pays off; they are still here. Finding deerflies near water makes perfect sense, as ponds are especially important deerfly habitats. As is true for other tabanids, deerfly larvae prey on aquatic invertebrates. They complete their aquatic phase as pupae before emerging as adults.

Both genders consume nectar and pollen, but only the females enrich their diet with blood. Whether the males of the species lack initiative to bite mammals we can't guess, but they certainly lack the equipment. The female's sharp bladelike mouth parts inflict painful wounds that make mosquito bites look positively genteel.

Biting flies elicit questions like, "What good are they?" Or more thoughtfully, "What is their role in nature?" And also, "Could we get rid of just this one species?" The disconcerting answer to the latter question is "yes," molecular biologists have discovered how to eliminate a species by inserting harmful genes that can be spread through an entire population. Although we have accidentally driven many species extinct, to my knowledge, the only deliberate extinction thus far has been smallpox.

Having discussed the important role that insects play in an ecosystem's food web and satisfied ourselves that driving deerflies from the planet was beyond our purview, my students and I resorted to a more local and fiendishly satisfying solution. We bought deerfly patches: double-sided sticky pads worn on our hats. When deerflies choose one of us as their next meal ticket they search for exposed skin. Does a deerfly patch look like human skin? You'll have to ask a deerfly. I won't question why they land on the patch, but I will take this opportunity to thank each and every one of them that takes that one-way trip and ceases orbiting my head.

To test drive the patch I parked near a campus pond. A deerfly landed on the side mirror—game on! Typically, I'd be swarmed in the field and at least one deerfly "guest" would join me for the car ride home. But this day would not be typical. I came forearmed. I had read the reviews and gawked in amazement at the online photographs of patches coated with innumerable flies stuck like so many dire wolves in a tar pit.

I emerged from the car, hat and patch on my head, and took a fifteen-minute walk between several ponds. During my walk I received one deerfly bite and swept another off my neck. I felt the familiar thuds of flies hitting my hat, but less orbital annoyance, it seemed to me. Wishful thinking? Time would tell.

The moment of truth: safely in my metal and glass cocoon, I removed the hat. Sure enough, the patch was emblazoned with fifteen deerflies, a single stray mosquito . . . and no gophers. I rarely endorse products, and indeed a good friend tells me that a loop of duct tape is just as good. Whatever solution you choose, at least deerflies need not force you to choose the indoors.

THE NORTHEAST'S MOST
ALARMING INSECT?

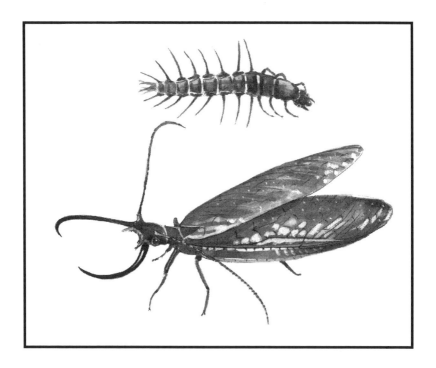

If freshwater insects did senior superlatives before graduating from
aquatic life, what would yearbook entries say about dobsonflies?
Largest? Most ferocious? Most likely to change names? Most likely
to bite a human? To be used as fish bait? Or to be confused with a
centipede?

All of these superlatives apply to larval hellgrammites—insects that,
upon emerging from the water, promptly change names to become
dobsonflies. These fascinating predators spend their larval stage eating
other invertebrates, including other hellgrammites. They're equipped
with impressive mandibles that can open wider than the width of their
own heads and can handily crunch through the tough exoskeletons of

most insects. An occasional angler has learned the hard way that the mandibles of larger hellgrammites are also quite capable of penetrating human skin.

Hellgrammites are important links in the food web between small invertebrates and fish. Six clawed legs, in addition to four hooks on prolegs at the back end, allow hellgrammites to forage over and under river rocks without being washed away. And they can be large. Topping out at 3.5 inches, hellgrammites are rivaled only by the giant water bugs for the title of largest aquatic insect on the block.

The paired lateral appendages on each abdominal segment may be the most intriguing features of these insects. At first glance they look like legs, which explains why my Saint Michael's College students sometimes think they've found some strange aquatic centipede. But these appendages are hauled along sticking out sideways and do not appear to help with locomotion.

So, what are these structures? Voshell's *Guide to Common Freshwater Invertebrates of North America* tells us they are gills. However, many hellgrammite species are well equipped with tufts of gill filaments that wave to circulate fresh water, while these lateral appendages do not appear to actively move. Researchers E. D. Neunzig and H. H. Baker, writing in 1991, suggested that the appendages are tactile, and that may well be the question addressed in a future experiment.

Adult male dobsonflies can reach 5.5 inches; roughly an inch and a half of that length is mandible. In late July and August, male dobsonflies account for most of my "alarming insect" identifications at Saint Michael's College and on social media. When summer visitors show up clutching scrambling captives in take-out food containers, I dramatically close my eyes and guess "dobsonfly"; often I'm right! The male's mandibles look like paired sickles, though thankfully, they're incapable of delivering a bite. The females, on the other hand, retain their biting function from the larval stage and should be handled with care.

What are those enormous male mandibles for? Whenever traits differ between genders, it typically has something to do with sex. The extravagant peacock tail attracts peahens; large deer antlers intimidate other males or are used to fight rivals. It seems that the latter situation applies to dobsonflies. Thomas Simonsen and colleagues from the

University of Alberta published photographs of male dobsonflies grappling and shoving with their mandibles. Eventually one managed to slide a mandible under his rival, and with a quick flip of the head, launched him off into the night. The one remaining male then devoted his attentions to a female and rested his mandibles across her wings. After some initial aggression, the female tolerated his attentions . . . but only to a point. She lost interest and he shuffled off rejected into the night; no baby hellgrammites were made that evening, at least not by the couple being studied.

When mating is successful, female dobsonflies lay their eggs on trees near rivers and other water bodies. After eggs are laid, the female coats them with clear liquid that dries to a chalky white and protects the eggs from drying out. Hatchling larvae crawl, or simply drop, back into the water and the next generation begins. Hellgrammites may spend as long as five years before emerging to pupate and hatch into the adult form.

And some of those adults emerging from Vermont's Winooski River make their way uphill to the Saint Michael's College campus, causing alarm for yet another generation of summer students.

THE CURIOUS CASE OF THE CUTE "FACE" CRANE FLY

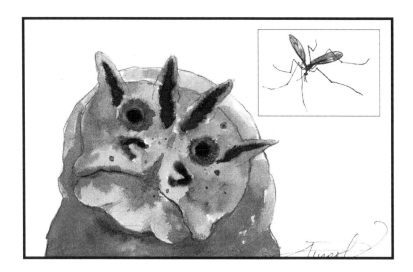

An email chirped in my inbox; "Check out the cute face on this insect we found." I opened the attachment (yes, from a reliable source). My colleague Professor Peter Hope had taken a spectacular photograph through his microscope. The larva in question had fallen into a pit trap set by our first-year Saint Michael's College students in Camp Johnson in Colchester.

The "face" seemed to have two very circular black eyes, a downturned smile, and a wild cartoonish hairstyle sprouting from lobes radiating in six directions. My esteemed colleague, a gifted botanist, had photographed the rear end of a crane fly larva. In fairness, any reasonable person might have made this mistake, especially because the front of the insect doesn't look like a front, its head pulled so far back into the body as to be invisible.

More than 15,000 crane fly species make up the superfamily *Tipuloidea*. They are often called "daddy long leg flies" because the larger species, with three-inch wingspans, sport spectacularly long legs. When I'm asked to identify a "huge mosquito," the answer is usually "crane fly." Smaller species, as little as an eighth of an inch in length, more easily escape notice.

Speaking of mosquitoes, crane flies are often called "mosquito hawks" or "mosquito eaters," even though the adults, if they feed at all, certainly do not chase down and eat mosquitoes.

Summer-flying adult crane flies are fascinating, but much more important biology happens during larval stages, which can be as short as six weeks or as long as five years depending on the species. Aquatic larvae continue to grow throughout the winter, feasting away in cold temperatures that put many of their fish predators off their supper; or in a state of torpor.

Students collecting river and stream samples are always impressed when their nets yield insects the size and shape of pinky fingers. Innards visible through translucent skin add to the fascination, or to the ick factor, depending on the student's viewpoint. These are the larvae of large crane fly species in the genus *Tipula*, and they are often found in streambeds, where they consume submerged leaves. Smaller crane flies, in the genus *Antocha*, fasten silk homes to submerged rocks and are far less conspicuous. Although small, they gather and consume large quantities of organic debris, and collectively can help to improve water quality. Regardless of their food source, crane flies in or near water risk becoming fish food and are of interest to anglers who, as described in Thomas Ames's book *Fishbugs*, tie "gangle-leg" flies to mimic the adult form.

Fish are by no means the only predators pursuing crane flies. Amphibians and reptiles also partake of the tipulid feast. A study in New Hampshire revealed that the adults are common menu items for little brown bats. Barn swallows and other birds also frequently snack on these insects. Crane flies are among the largest insects eaten by some species of swift (first cousins of swallows), making them a most valuable prey item during the nesting season.

Although crane fly larvae are best known as aquatic insects, there are also terrestrial species that occupy habitats from tundra to desert. For example, a type of crane fly larvae dubbed "leatherjackets" can cause yellowing and bald patches in European lawns by devouring both roots and grass blades.

Unfortunately, two European leatherjacket species have been detected in central New York and in Long Island and may well munch their way through the Northeast in coming years. These new pests have also been in Ontario since the 1990s. According to Pam Charbonneau of the Ontario Ministry of Agriculture, Food, and Rural Affairs, starlings and skunks do additional lawn damage in search of the juicy little moveable feasts. Large numbers of the larvae are sometimes forced to the surface after heavy rain and might be a gardener's first clue to the cause of their yellowed lawn.

Regardless of habitat or food habits, crane fly larvae tend to have distinctive "facial" features on their rear ends. (To help students instantly identify the larvae, a grad school colleague liked to say, "Tipulids have traces of faces round their anus.") The dark "eyes" are in fact spiracles, or the openings to the insect's respiratory system. The mane-like lobes that surround the spiracles, or form a crown shape, are prehensile in some species and used for movement and other functions. One aquatic species uses water-repellent hairs on its lobes to contain a buoyant cup of air that suspends the larva from the water surface while also facilitating respiration.

While preparing an insect ID cheat sheet, I recently asked Professor Hope to resend me the photograph as an example for students. This time the subject line read "Fly butt photograph." It seems that Peter is learning his insects faster than I'm learning my plants.

BURROWING MAYFLIES

W hen I visit Lake Champlain's waterfront during summer, I often see slim insects with sail-like wings, perched on trees and clinging to window screens. These are burrowing mayflies, from the family *Ephemeridae*.

The "burrowing" part of these insects' name relates to their early, aquatic stage of development. As nymphs, they live in U-shaped burrows on the lake floor. By undulating their gills, the nymphs pump water in one end of the burrow and out the other. In this way, they harvest oxygen and tiny food particulates from above, and pump out carbon dioxide, frass (insect poop), and other waste.

The volume of water moved by a single nymph is miniscule on the scale of a large lake, but there can be hundreds of thousands or even *millions* of nymphs in a single body of water. Collectively, burrowing mayfly nymphs pump enormous water volumes through lake-floor sediment.

Entomologist André Bachteram and his colleagues at the University of Windsor ran experiments with mayfly nymphs and determined that as food is depleted, the rate of pumping increases. The significance of this finding is that it demonstrates that the nymphs' pumping behavior can counteract the water-clearing activities of invasive zebra mussels. By pushing fine sediment particles back up into the water and increasing this activity in response to less food-rich water, nymphs contribute to a lake's state of "bioturbation," in other words how many nutrients and other sedimentary materials are resuspended in the water by living organisms, and made available for the pelagic food web. Bachteram determined that hungry mayflies in parts of Lake Erie push sediment particles out faster than zebra mussels could remove them.

Burrowing mayfly eggs sit for a few weeks to months before the tiny nymphs hatch. The nymphs spend nearly two years growing and molting up to thirty times, before one final molt liberates a nearly adult stage that is unique to mayflies. They emerge from the water as a winged subimago or "dun" stage, and fly onto land, where they once again molt into a glossier and less hairy reproductive adult, or "spinner." Only when they attain this final stage can they mate.

Emergence is tightly synchronized and en masse. There are at least a couple of compelling reasons for this flash mob approach. In their final two airborne stages, mayflies are short-lived; they don't feed or even have mouthparts. So it's advantageous to emerge in proximity to masses of mayflies of the opposite gender. Another reason is that those mayflies that end up as fish food don't reproduce; a mass emergence overwhelms predators and betters the individual odds for survival.

And so, all at once, a great many mayflies dig out of the muck and swim for it. Consider this phenomenon from a fish's viewpoint. Nymphs that have spent two years tucked inaccessibly away in muddy burrows suddenly appear as a moveable feast. Organic particles that were useless to the average fish have been methodically filtered from the water over long months and reassembled into fattened, high-protein snacks now rising through the water, dallying briefly on the surface to ditch their last nymphal exoskeletons, and then taking flight across the waves.

These enormous hatches turn the heads of hungry salmon and trout that soon turn the heads of anglers. Fish and other predators do something called "prey switching"; they gorge on the most common available prey; until different prey become more common and then they switch again. Experienced fly fishers are well aware of this fishy behavior and so they "match the hatch" when selecting flies to cast. If trout have switched to mayflies, then even the most delectable caddisfly is unlikely to attract a nibble.

Lake Erie mayfly hatches were once so massive that shopkeepers used snow shovels to clear sidewalks. Reports from the 1940s describe buildings blackened, trees so laden that branches bent and broke, and electrical fires caused when mayflies shorted out neon lights that attracted them in droves. However, organic pollution, and excessive loading of sediment and nutrients that sucked most of the oxygen from the lake floor brought the huge hatches to a halt in the 1950s.

In the 1970s, the United States and Canada enacted environmental protections to clean Lake Erie. Effects were noticed by the mid-1990s, when biologists found small burrowing mayfly populations—at first near shore and later out in the broader lake. Mayfly swarms have since recovered to such a scale that they are sometimes detected on weather radar and recently produced the headline "Swarms of Mayflies Invade Cleveland."

Mayflies have not returned everywhere or to the pre-pollution levels, but recovering populations at any level give me hope. More importantly, these recoveries demonstrate that patient, concerted environmental improvements that are informed by science can and do work.

WATER DRAGONS

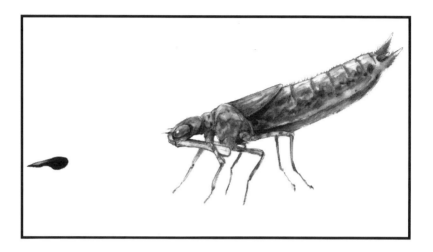

The same dragonflies that buzz our summer picnics spend their youth as aquatic nymphs, stalking or waiting for prey in streams, lakes, and even tree holes left where branches have snapped and rotted back. Multiple generations live side by side in the same pond; youngsters with months or even years of aquatic life ahead, and their elders, soon to emerge from the water as adult flyers.

An adult dragonfly does nearly everything on the wing. It uses all six legs to form a catch net, scooping prey right out of the sky. Dragonfly nymphs, however, use their legs strictly for walking. They have no need to grapple prey with their legs because they, along with their cousins the damselflies, have special appendages called masks that are lost in the transition to the adult stage.

The mask is located beneath the head and projects back between the front and middle pairs of legs. As the name implies, it also covers part of the dragonfly's face. To visualize this appendage, hold your two hands flat, side by side with palms facing you so that your pinkie fingers touch along their lengths. Now bring your elbows together and cover your mouth with your hands. Your shoulders are analogous to the joint that attaches the dragonfly's mask to the underside of its head; your elbows represent a second joint that allows for extension.

Now, keeping your elbows and hands together, reach for imaginary prey over your head—perhaps a juicy deerfly larva? Congratulations, you have approximated the hunting technique of a dragonfly nymph, while possibly adding a new yoga move to your practice! Except that the insect's movement is dramatically faster than any human could muster.

Some dragonfly nymphs stalk prey to get within striking distance for their masks; others simply wait. The *Gomphidae* and *Aeshnidae* dragonfly families grow a fairly flat mask with opposing hooks at the end that grasp prey; members of the *Cordulegastridae* and *Libellulidae* families have spoon-shape masks with a split down the front like a zipper. These open and close around the catch of the day.

Either way, the prey is rapidly drawn back to the jaws, which cut it into bite-sized chunks for swallowing. Little is wasted; leftover scraps tend to accumulate on the mask and are consumed as a final course.

All of this stalking and hunting requires oxygen, and aquatic organisms have diverse ways to extract oxygen from water. Some shark species, famously the great white, must swim to breathe; most fish can pump water over their gills and therefore can be still. Dragonfly nymphs also pump water over their gills, but their gills are not up front like those of fish. Fresh water is sucked in through an opening in the dragonfly nymphs' rear ends to bathe gills arrayed inside their abdomens.

Brian Swisher, a colleague at Saint Michael's College, recently showed me some newly hatched nymphs; the baby dragons, scarcely as big around as pencil lead, were so translucent that their respiration was visible under a microscope. I could see the three protective plates that surrounded the respiratory opening, closing and opening as the water was inhaled and exhaled. Minute hairs on the edges of these plates serve as sieves, excluding particles that might otherwise foul the gills.

If an organism is going to commit to the indignity of breathing out of its hindquarters, it may as well take full advantage of this superpower. Dragonfly nymphs can use their respiratory holes for rapid water ejection, blasting themselves away from harm. To see this in action, place a nymph in a basin of water from its pond or stream, with a little sand at the bottom. Gently poke it. Usually, the nymph will move away rapidly, and you will see sand being washed aside by the force of its jet.

Dragonfly nymphs aren't picky eaters. Their prey includes other invertebrates and even small fish and amphibians. I am sometimes asked: What good is a mosquito? At the very least, mosquitoes make wonderful food for dragonflies, for both adults and nymphs. In a world in which we have the technology to deliberately eliminate species, it's worthwhile to point out that even the seemingly most superfluous creatures have relevance in the food web. I like Aldo Leopold's response to similar questions: "To keep every cog and wheel is the first precaution of intelligent tinkering."

THE CASE OF THE
CONFUSED KINGFISHER

In July 2019, Rich Kelley posted a most unusual photograph to the
Vermont Birding Facebook group with the caption "Someone bit off
more than he could chew . . ." The photo, taken in the Missisquoi
National Wildlife Refuge, showed a belted kingfisher weighted down by
a mussel clamped firmly onto its beak. They were locked in an embrace
that, absent intervention, would have been fatal for both; thankfully,
Rich effected a rescue.

For me, the photo inspired a rare eureka moment. I strung a few
scientific facts together and jumped to a possibly outlandish conclusion.
This had been deliberate action by the mussel, doing what mussels do,
misinterpreted by a kingfisher, doing exactly what kingfishers do.

Before you write me off as a biologist out in the sun too long, bear
with me while I present the facts. First, consider the kingfisher. The
name says it all: they fish and fish well. Kingfishers are visual predators
that drop from high perches to execute full-body dives, so their spear-
shaped beaks cut through the water but not through the fish! That's
right, contrary to popular belief, kingfishers don't spear fish. Instead,
they open wide and close their bills to firmly grip their slimy prey.

To really evaluate my speculation, I needed to find out if the belted
kingfisher menu extends beyond fish to include, perhaps, eastern lamp

mussels. I learned that the kingfisher's diet may comprise insects, cray-fish, and occasionally small mammals, amphibians, and reptiles. But the king's share of their diet consists of fish. Nowhere in the papers I consulted were mussels of any type part of the kingfisher's meal plan.

The first part of my quest was complete: it seemed unlikely that the kingfisher in question was out hunting mussels. My next step was to explore mussel biology to see if the timing of Kelley's photo was right for my hypothesis.

Eastern lamp mussels don't generally travel far. To disperse, they act like submerged hitchhikers. Larval lamp mussels, called "glochidia," attach to fish gills for about a month and travel as far as the fish swims before dropping off in their long-term home. How they attach to the fish is the link in the chain that may just have accidentally ensnared a Vermont kingfisher.

Lamp mussel glochidia don't swim, and they remain in their mother's pouch until a fish makes direct contact with mother mussel. Direct fish-mussel contact has not been left to chance. When the season is right, the mother lamp mussel grows an extraordinary structure that hungry fish just can't resist.

Sprouting from its soft tissue, and protruding between the half shells, the mussel grows a near perfect "fish." At the "head" end is often an eye-spot, and at the opposite end, a tail complete with fishlike patterns and even a delicate set of fins. To complete the ruse, the mussel twitches its little lure, like any good angler, to catch the eye of a passing fish. Ideally, the lure attracts a yellow perch, which is the only known host for eastern lamp mussels. Real fish strike at the little fake "fish," and the mother mussel ejects her glochidia, which promptly clamp onto the fish's gills.

Another question essential to my theorized plot is when does all of this happen? Scientists on Lake Ontario found glochidia on yellow perch gills in both May and August. So, it's reasonable to expect appetizing mussel-fish lures in July in Lake Champlain.

If the mussel lures are convincing enough to entice fish to strike, it's a small leap to imagine that they may fool a different visual fish predator: the kingfisher. A leap of faith and imaginative conjecture, however, is not enough to provide scientific certainty. To truly support my hypothesis, I'll need to schedule an interview with a certain kingfisher. And I'll ask the question every teenager dreads: "What WERE you thinking?"

PART VI

WHAT LIES AHEAD

*Do unto those downstream as you would
have those upstream do unto you.*
—WENDELL BERRY

Each of us standing on dry land is standing in a watershed and our choices impact the human and nonhuman organisms downstream. At the same time, unless you are on a mountaintop, you too are downstream.

Society as a whole in each watershed determines whether the Chesapeake Bay will support a blue crab fishery, Lake Champlain will have toxic algal blooms, the Colorado River will contain water when it reaches Mexico, or the Cuyahoga River will catch fire on a regular basis.

Each of us is a member of a nested series of societies from household to neighborhood, town, county, state, and nation. Unlike global carbon pollution, local landscape improvements have local incremental positive impacts on aquatic habitats. But in common with broader environmental problems, these local improvements sum downstream to improve larger scale conditions.

So, what can we each do? And who am I to tell anyone what to do? Perhaps if I tell you some of the things that I am doing it will illustrate some solutions. My roof water that previously drained to the driveway is now piped into a garden. When developers claimed that a wooded area of my town was unimportant to wildlife and should be razed for housing, I installed trail cameras to show bobcats, gray fox, raccoons, and otters using the area and wrote about it for my local paper. I lobbied the college administrators and trustees to restore floodplain cornfields to floodplain forest and now I collaborate with other college employees and students to plant trees there.

What you can do to promote sustainable behaviors depends on your particular circumstances, but each of us has a contribution to make and a responsibility to make it. It is my hope that this book introducing some of the amazing life in fresh water will foster appreciation and in turn, appreciation will foster care. This last part of the book covers some of our current problems and offers solutions for what I am optimistic will be a better future. After all, there are still blue crabs in the Chesapeake, and the Cuyahoga last caught fire in 1969. Change is possible!

WHAT LIES AHEAD

It would be tempting to chronicle the injurious assaults humankind has launched against our life-sustaining waterways. And indeed such an exercise has value. Instead, in this part of the book, I have endeavored to focus on solutions, both large and small, enacted by individuals and governments to protect and improve the integrity of freshwater ecosystems.

It is a simple fact that we require fresh water for life, and it is undeniable that we are the single species that does most damage to our own water supplies and therefore to the life found in fresh water. But we also have the knowledge and ability to address these challenges, and in many cases are doing so. Freshwater science has matured to a point where concrete recommendations can be made to protect lakes, rivers, and streams that in turn provide for and protect us.

Eutrophication caused by nutrient pollution is the most common impairment of fresh water in most places. We simply produce and release more nitrogen, and particularly phosphorus, into fresh water than can be sustained. The results of nutrient pollution include algal blooms, waterways choked by vegetation, toxic releases from blue-green algae, and oxygen deficits that kill fish and other life-forms.

Pointing a finger at sewage plants, golf courses, urbanization, and agricultural conglomerates is disingenuous because each of these nutrient sources is driven by human demands, by our demands. But identifying the sources of problems is the essential first step toward fixing them. And so some essays in this part of the book focus on the root causes of eutrophication and on solutions.

Climate change is without a doubt the single greatest challenge we currently face. The signal of this change has been recorded by trees advancing up Vermont mountains, extended growing seasons, and reduced ice coverage in lakes and ponds. Climate influence on aquatic systems is not limited to melting ice; increased rainfall in the Northeast for example has increased erosion, exacerbating our eutrophication issues.

Other freshwater impacts such as dams and restricting rivers to narrow channels are widespread, but the solutions are local in nature, applied to one stream or river at a time. This part of the book includes essays on dam removal and using fallen trees to restore rivers and streams to more natural states that improve water and nutrient retention.

Finally, we have moved exotic species from continent to continent, often with disastrous results. Fresh waters have been invaded globally with American species moved to Europe, Asian species moved to the Americas, and Australia invaded by a toad native to Central and South America. Shamefully, the case of the cane toad in Australia was caused by misguided science. The final essay in this part of the book covers invasive zebra mussels, just one of more than 2,500 exotic species recorded from the Laurentian Great Lakes.

My intent in writing the essays in this part of the book is to point out that we can make positive changes to improve our freshwater assets. Restoration ecologists point out that the name of their field is impossibly ambitious; "rehabilitation ecology" might be better. Suggesting complete restoration of dramatically degraded ecosystems as an end goal would be as naive as suggesting complete restoration of perfect health following decades of smoking, but what sort of doctor would not at least suggest some rehabilitation?

So I'd like to advocate for rehabilitation. Let's remove dams where we can, add and protect riparian vegetation, daylight urban streams, and control urban runoff. We won't return to some utopian ideal, but we have the power to make positive changes. Let's do that!

KEEPING IT CLEAN
DOWNSTREAM

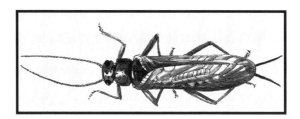

In peaceful streams, aquatic macroinvertebrates such as crayfish, stoneflies, and caddisflies travel over and under submerged rocks, foraging for leaves, algae, and other invertebrates. When rain falls, their world turns upside down. At first only the surface is disturbed, but before long, runoff reaches the stream and increases its flow manyfold. Silt and sand blast every exposed rock surface. At peak flow, boulders are propelled downstream by powerful currents.

How do small creatures survive such crushing chaos? They hunker down. Water-filled nooks and crannies extend deep below streambeds and far beyond riverbanks in an area called the "hyporheic zone." These deep interstices provide a safe haven even while turbulent water pulverizes the riverbed, comparable to a storm cellar in a tornado.

Stormwater has great destructive potential, but it also flushes and cleans aquatic habitats. Riverbeds are rejuvenated when sediment is flushed from the bottom and deposited on floodplains. Algae and bacteria grow back rapidly on the scoured rocks. Macroinvertebrates climb out of their cellars and return to foraging. The cycle of storm, recovery, and regrowth is the natural state.

You can see the effects of this cycle yourself by looking at river rocks. In a healthy stream, you'll find rocks perched on other rocks, with the streambed visible under the water, and little silting. Although, just after a storm, the rocks may be scoured clean, they'll soon develop a slimy covering of algae and host a diversity of tiny creatures.

In polluted streams, however, you're likely to find something very different. River rocks may be embedded in silt, and when you pick them up, you'll find that they're wearing bathtub rings, with algae only growing on the upper half; or in particularly impacted streams, the upper third or quarter. Stream ecologists use such observations of embeddedness as an index of stream health.

Poorly planned development disrupts the cycle of streambed renewal. Where stream banks are bare, erosion can be a big problem. Soil lost from overgrazed or overcropped land ends up in the water, where it plugs the streambed nooks and crannies. Imagine a concrete truck unloading through your window and filling your home with a solidifying mess. Only the hardiest of invertebrates survive these conditions, and the whole riverine food chain can be affected.

Traditional paving and buildings also create problems, as impervious surfaces dramatically increase the volume of water sent straight to streams. Formerly small, cool, perennial streams can become torrents of unnaturally warm water. Channels become deeper and eroded materials are deposited in streambeds. Since rain doesn't reach the ground underneath the pavement, groundwater can become depleted, and the streams may run dry between rainstorms.

Farming or urbanization won't disappear, but there are ways to intelligently develop landscapes to better protect streams. For example, Saint Michael's College has installed a system of curbed parking lots connected to rain gardens. These are shallow, gravel-lined depressions, strategically planted with vegetation that tolerates occasional submersion. The rain gardens easily absorb water from typical rainstorms and, together with the dish-shaped parking lots, can even contain all of the water dropped during one-hundred-year storm events. Saint Michael's has also replaced many impervious sidewalks with attractive, pervious, bricked footpaths. Roof water from the gymnasium runs into deep gravel beds. Runoff from recently constructed roads collects in an underground tank. All of these systems drain gradually into the ground, drastically reducing the downstream potential for erosion. The recharged groundwater keeps a small perennial stream flowing to the Winooski River even during dry spells.

There is no doubt that people affect stream macroinvertebrates and the fish that they sustain. We do, however, have the choice to protect our streams by thoughtfully managing our impacts and reducing erosion in urban and agricultural settings. With respect to Joni Mitchell, there's a lot of room for ingenuity between paradise and a paved parking lot.

WETLANDS REDUCE
FLOODING AND IMPROVE
WATER QUALITY

I grabbed my camera and visited the Saint Michael's College Natural Area following some heavy rain late this spring. My goal was to capture exciting photographs of rushing Winooski River waters. I was to be sadly disappointed!

Floods on television are dramatic affairs. Rising rivers rush violently down streets carrying away possessions and leaving tragedy in their wakes. The news leaves the impression that floods are aggressive events. And indeed, during my frequent Natural Area walks, there is evidence of large movement of soil, debris, tree limbs, and even whole trees. An oxbow provides evidence of a substantial change in river course and aerial photographs from the 1930s confirm that the oxbow is precisely where an earlier Winooski River channel flowed.

Following one flood last year, more than a foot of soil was deposited up to the base of a trail camera I had foolishly installed in the floodplain. While digging the camera straps out of the mud, my eye was drawn to a tree trunk lodged in the fork of a sapling. The weight had bent the young tree over, splitting the bark near the base.

But my spring flood was a gradual event with little in the way of rushing water. Rather, there was an imperceptibly slow rise in the water level. The most obvious evidence of flooding was that previously high and dry locations were now submerged.

Other clues emerged as I walked farther. Streams of bubbles rose from freshly flooded soil as each and every worm or insect burrow gave up its air to the advancing water. Live grass underwater was a definite clue; few grasses like to be submerged and don't grow in active river channels.

I followed the flood line as it staggered and looped across the landscape relentlessly, patiently finding all low-lying land. Where the river reached the lip of low-lying depressions, I witnessed the overflow turn depressions into ponds that would outlast the flood. Even these overflows lacked the drama that would make for good TV. I marked them with my GPS and checked them three days later. Sure enough as the river receded, the ponds had retained floodwater, providing valuable habitat for as long as they lasted.

My return trip painted an interesting picture of the ghost of the flood now past. Every leaf, grass blade, and stalk in briefly submerged areas had been coated thoroughly in fine silt. Nutrient-rich soil eroded from upstream and bound for Lake Champlain had been intercepted and stored by these floodplain wetlands. The simple existence of this wild place on a suburban campus was reducing eutrophication and algal blooms by cutting off the food supply that would otherwise foul the lake. Silt deposited on the plants would be washed down to the soil by spring rains to feed vegetation at the base of the riparian food web. The flooding provided a necessary replenishment of the wetland ecosystem, and the wetland in return was reducing Lake Champlain beach closings.

And my walkabout provided a replenishment of my psyche and a revamping of my teaching approach that all the stale scientific journals on the planet could not rival. Much can be learned simply by walking and watching; my students will do the same. For years, I have been teaching that wetlands are the kidneys of our landscapes. They filter nutrients and pollutants and convert them to plant material feeding other organisms, etc. etc. My simple stroll confirmed what this looks like on the landscape. The coating of the plants was a vivid detail that had not occurred to me to include in my lectures and did not merit a mention in the textbooks.

In my simplistic textbook view of riparian flooding, I assumed that nutrient reduction was accomplished by settling out of materials into pond-like wetlands. But if that was the case, all we would need to control nutrient pollution would be some simple concrete basins. Floodwater in; silt accumulated; pumped out; repeat. Indeed, that approach is implemented in many places, although the necessary pumping is too often neglected.

But a true natural wetland, or a constructed wetland like one I frequent in Fort Ethan Allen in Colchester, Vermont, has far more going on than could occur in a very large steep-sided concrete bathtub. Riparian wetlands capture floodwaters, providing an overflow mechanism for rivers in flood, and evaporate or slow-release that water back to the river.

In August 2011, Tropical Storm Irene dropped more than 10 inches of rain in parts of Vermont onto already saturated soils. It caused massive destruction across the state and also provided a graphic example of the flood-control function of wetlands. When the floodwaters reached Rutland, Vermont, Otter Creek was walled up between concrete barriers with no place to go but up and over those barriers with devastating results. That would be the place to send the TV cameras, and indeed photographs of that catastrophic event graced our newspapers for weeks.

As Otter Creek flowed downstream, it continued to accumulate rainwater from the stalled storm. The river swelled and grew larger. Intuition would suggest that the town of Middlebury, downstream of Rutland, would have been laid waste by the rising flood. But Middlebury was largely spared from wholesale destruction. No mysterious force prevailed upon the waters; there was no intervention divine or otherwise. The water simply spread out into the natural wetlands of Addison County and on into fields enriching the soils as floodwaters have for millennia.

There was a time when our national policy essentially stated that the only good wetlands were those that had been drained, filled, and built upon. Hopefully we have moved on from that dark phase of our history and wetlands have been given many of the protections they deserve. And they protect us and improve our water quality in return. There's more to learn; more wetlands to protect, restore, and create *de novo*. Climate change has increased flooding in the Northeast. With forethought, more of our future floods can be quiet affairs to watch with fascination than to flee from in fear.

MALARIA MOVES NORTH, INFECTING LOONS

(Co-authored with Ellen Martinsen)

W ith its striking black-and-white coloration and haunting calls that echo across the water, the common loon is an iconic species of the Northeast. Once widespread and truly "common," loon populations have been diminished by lakefront development, acid rain, mercury deposition, lead fishing tackle, watercraft collisions, and environmental toxins. While dedicated conservation efforts in recent years have helped loons recover, populations are not yet stable, and studies by Kristin Bianchini of Birds Canada and Walter Piper of Chapman University predict steep declines in coming decades. An emerging malaria epidemic may compound the problem.

Malaria affects humans in the tropics, but this debilitating disease can also harm many other species, including other primates, white-tailed deer, bats, lizards, and several bird species. Historically, loons have been exempt. But a warming climate and subsequent expansion northward of several species of birds—and the parasites they carry with them—is proving deadly for loons.

A 1947 edition of Peterson's *A Field Guide to Birds* indicated that tufted titmice ranged as far north as New Jersey, northern cardinals made it to southern New York, and common grackles occurred up to southern New England. These birds are now common in northern New England and New York, and the list of birds expanding northward is growing.

Negative impacts of songbirds on waterbirds may seem unlikely—loons do not compete for seed at bird feeders, and cardinals have not developed appetites for yellow perch. However, as songbirds extended their range, they brought malaria parasites along with them. Expanding insect populations—namely mosquitoes—exacerbate the problem.

Mosquitoes transfer malaria parasites from bird to bird. So an increase in the number—and type—of mosquitoes can mean a potential increase of malaria transmission. Tanya Petruff and Joseph McMillan of the Connecticut Agricultural Experiment Station and their collaborators measured a 60 percent increase in mosquito numbers in Connecticut between 2001 and 2020 and 10 percent more mosquito species in the state. That's a lot more biting, bloodsucking, and potential for transfer of disease.

Of particular concern for loons is *Culex erraticus*, a mosquito previously restricted to southeastern states. Starting in the 1980s they were detected in New Jersey and have been migrating north ever since. They are now found throughout New England with the exception or Maine. Andrea Egizi from Rutgers University and her colleagues sequenced DNA from blood consumed by this species, which revealed that the most dined-upon waterbirds, included loons. And *Culex pipiens*, the house mosquito introduced from Europe—now abundant in our backyards—transmits a malaria parasite species also recently detected in loons.

Thus far, six malaria species have been found in loons using PCR, the technique used to detect COVID-19 in humans. It is unclear if all six species harm loons, but volunteer-driven fieldwork, necropsies by veterinary pathologists, and molecular biology indicate that at least one species of malaria parasite certainly kills loons.

The story first unfolded in 2015, when paddlers on Lake Umbagog—which spans northern New Hampshire's border with Maine—reported a deceased loon. Wildlife biologists brought the loon to Inga Sidor, veterinary pathologist at the University of New Hampshire, whose examination revealed a grossly enlarged spongy heart, degraded lungs, and a greatly enlarged spleen. Microscopy of tissues revealed malaria parasites clogging the capillaries supplying blood to the brain. The cause of death was cerebral paralysis. Cerebral malaria affects more than half a million people in sub-Saharan Africa annually, and the parasite occurs globally in tropical and subtropical climates. But malaria had not previously been found in loons.

Ten more loons have since been found to have died from malaria in Northern New England. Detective work has revealed that all ten were infected by the same malaria species as the Umbagog loon. This deadly malaria parasite, identified by DNA sequence data, can apparently easily jump from one bird host species to another.

Pathogens and hosts typically coevolve to mutual tolerance, and it is not in the best interest of a parasite to be lethal. Dead hosts don't transmit pathogens, since the pathogen dies with the host before reproducing and spreading its genes. Pathogens thus evolve reduced virulence or detrimental impact on their precious host organisms.

But when pathogens move to new hosts, problems arise quickly, including disease and death. HIV, a benign chimpanzee virus, became lethal when it moved to humans. Similarly, a malaria parasite species that is harmless in northern cardinals, tufted titmice, or other unknown hosts could be lethal in loons. Interestingly, the deadly malaria parasite is common in the American robins that love our lawns.

This all begs the question: What is the human malaria risk under changing climates? Avian malaria parasites are strictly limited to birds; there is little risk that robins or loons will transmit malaria to humans.

Human malaria parasites are also on the move, but let's not rush to blame the birds.

Deceased loons should be reported to your state loon biologist, local loon non-profit organization, or state wildlife agency. Time is of the essence, as malaria becomes more difficult to detect as decomposition advances. Learn more about the research here: (https://wildlife pathogens.org/loon-malaria/)

ICE-OUT AND
CLIMATE CHANGE

W hile driving down from Isle La Motte in early December, my son and I noticed a fine skim of ice floating down Lake Champlain's Alburg Passage. As it collided with the Route 2 bridge supports, it broke into rectangular fragments. I wondered if what I was seeing was typical or a symptom of changing climate? But a single observation tells you only about the current weather and says nothing about climate trends.

To understand long-term patterns requires long-term data. So I reviewed ice formation data on Lake Champlain. I learned that between 1816 and 1916, the lake was "closed" to navigation in ninety-six out of one hundred winters. In the last thirty winters, the lake has closed thirteen times, and just three times this past decade. At first blush, this might seem like overwhelming evidence for less ice, but again, this is not the whole story.

The two-hundred-year dataset was gathered by three different governmental agencies, a Burlington public official and historian, and a "cooperative weather observer." Consistency might be a bit much to expect and "closed to navigation" could range from an ice passage from Vermont to New York, or simply frozen harbors.

For a more consistently measured dataset, Dr. Alan Betts, a Vermont climatologist, looked to the Joe's Pond Association. Each winter for more than two decades, association members have placed a wooden pallet on the ice of Joe's Pond in West Danville, Vermont. A cinderblock sits on the pallet and is strung to the plug of an electric clock on Homer Fitts' deck. For a small donation to the association, you too can guess when the ice will give way, the cinderblock sinks, and the clock will be unplugged; best guess wins!

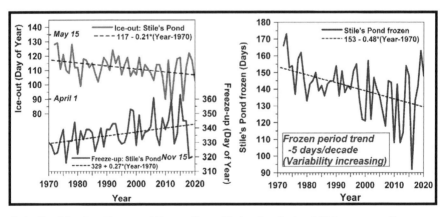

Stiles Pond Ice-Out Dates and Frozen Days, Updated to Spring 2020, courtesy of Dr. Alan Betts.

The simplicity and consistency of this measurement technique is precisely what piqued Betts's interest. Across the short interval of twenty years, there's a clear trend; the cinderblock sinks about six days earlier than it did two decades ago.

Betts has also reviewed forty years of ice-out data from the Fairbanks Museum and found the same pattern: the ice on Stiles Pond goes out about three days earlier per decade. Every decade, on average, the pond has frozen four days later, and the total frozen period has been shrinking by seven days per decade since 1970.

"Ice-out" patterns are consistent with other indicators of change. For example, Betts has reviewed data on Vermont's lilac flowering dates. On average, lilac leaves are developing about two weeks earlier than they did in the 1960s, and flowers open more than a week earlier.

With the greatest respect to ice and lilac, I suspect that some Vermonters might be more interested in changes in maple sugaring; this thought has not been wasted on Vermont scientists. Justin Guilbert and Vermont EPSCoR collaborators examined climate trends and predicted eleven fewer maple sugaring days by mid-century. They also predicted a shift in the sugaring season toward the midwinter months of December and January.

If at this point, you're thinking that these trends are awfully short-term, and that anyone trying to predict the future of sugaring is walking on thin ice, you have a valid point. One of the difficulties of predicting climate change and its effects is the complexity of factors, including the background "noise" of our naturally variable weather conditions. This is, after all, a region which prides itself on the notion that, "If you don't like the weather, wait a few minutes."

That said, while we can't with certainty predict what maple trees, or ponds, or ornamental plants will do in future years, it's very clear that we're in a period of rapid temperature change; and based on what we know of atmospheric science and human-caused emissions, there's no reason to expect that change to stop any time soon.

As Betts recently told me, "Climate change is on a roll and all we can do is slow it down, and give our societies and all of life on Earth more time to adapt." In the meantime, I plan on placing my first ever bet on Joe's Pond this year. What day will I bet on? That's my secret, but it will certainly be much earlier than I would have bet in 1997.

.

LET SLEEPING LOGS LIE

For more than a dozen summers, my Vermont EPSCoR colleagues and I lead high school student and teacher research projects. After training in July, research teams gather data from streams across Vermont and as far away as Boston and Puerto Rico. The following March, the teams return to Saint Michael's College to present their findings in a research symposium.

Along with macroinvertebrate samples, water samples, and flow measurements, our teams measure fallen logs in their streams. This past year, a student asked me why we care about logs. What difference could a dead piece of lumber make to stream health? These are important questions!

Every natural history fan knows that fallen trees on the forest floor are essential to insects and in turn to everything from woodpeckers to bears. While woodpeckers gorge on insects on dry land, river ecologists consider fallen logs, or "coarse woody debris" equally important to river health.

Rivers and streams are sometimes treated as mere drainage features; ways for rainfall to efficiently exit the landscape. Until recently, state and federal agencies managed our streams in exactly that way. Streams were straightened or "channelized," the banks armored with riprap, and many streams piped underground speeding water along, sometimes with disastrous downstream impacts. Fallen logs were obstructions to the plumbing, or hazards to navigation, and routinely removed.

Rapidly flowing water in straightened streams erodes streambeds sinking streams deeper into the landscape like a hot knife through butter. One such "incised" stream on the Saint Michael's College campus undercuts the culvert that pipes the stream under a road. At the downstream end, the water has washed out and deepened the streambed, creating a waterfall below the pipe that is impassable to migrating fish.

Straight and tidy streams are conveniently predictable, but are poorly received by fish, amphibians, and the invertebrates supporting aquatic food webs. Just as some birds favor treetops, while others frequent trunks or the forest floor, river organisms have their niches. A complete river or stream community has pool residents such as dragonflies, fast water organisms including riffle beetles and stoneflies, and backwater fauna such as water striders and whirligig beetles. Piped or channelized streams are carefully engineered so that water flows rapidly with little turbulence and few eddies. Elimination of several ecological niches is a direct consequence of this engineered approach to stream management.

Backwaters and mid-channel gravel bars that were eliminated in the name of efficiency also eliminate diverse flora and fauna. Fish requiring slow eddies, backwaters, pools, or rocky riffles simply can't survive in a flowing sewer. Monotonous predictability is a far cry from the natural state of streams where water slows with every twist and turn and accelerates through runs and riffles only to slow again in deep pools.

Logs and branches interrupt water flow and encourage natural meandering. Fallen leaves that accumulate on upstream surfaces of tree limbs are both essential habitat and the base of aquatic food webs. Trees that land "just so" can dam streams, forming substantial pools.

Channel-spanning dams are more common than one might expect because water frequently undercuts soil on the stream side of trees, causing them to fall toward the water. Branches and trunks in streambeds provide areas of slack water where macroinvertebrates can flourish and fish can hide, avoiding the torrential rush of the main channel that is occupied by still other organisms.

Woody debris slows stream velocity, causing gravel, sand, and silt deposition. In short order, a fallen tree causes sandbar formation forcing the water right or left, restoring the meandering structure eliminated by older management approaches. Stream restoration practitioners use this knowledge to heal damaged waterways. The Izaak Walton League's *Handbook for Stream Enhancement & Stewardship* encourages installation of felled evergreen trees against eroded banks in stream channels to prevent further erosion and to accumulate sediment. In *Let the Water Do the Work; Induced Meandering, an Evolving Method for Restoring Incised Channels*, Bill Zeedyk and Van Clothier describe picket baffles where wooden posts are installed in streambeds, again to accumulate sediments from flowing stream water and re-form meanders.

Sediment and organic material accumulated among branches on streambeds, stores nutrients that would otherwise move downstream and contribute to eutrophication. Wood in streams therefore helps to solve the eutrophication and harmful algal blooms plaguing our lakes.

Flood control is another issue where wood has a positive influence in streams. Counterintuitively, straightening or channelizing streams to control flooding can have opposite effects. We learn in grade school that a straight line is shorter than a wavy line between the same two points; by the same logic a meandering stream holds more water volume than a straight one. Meandering streams and their wetlands contain and slowly release water that a channelized stream would deliver directly to an unfortunate community downstream, causing far more damage to streambeds, banks, and infrastructure such as bridges.

Some invertebrates use submerged wood directly. More than twenty different aquatic insect species in North America require wood for survival. These so-called miners tunnel into underwater logs and branches consuming wood as they go. The miners include some beetles, mayflies, caddisflies, stoneflies, and true flies. These invertebrates in their turn contribute to a diverse food web upon which fish, birds, and mammals, including us, depend.

And so, whether you are managing a woodlot, caring for the ditches on a family farm, or participating in a stream cleanup, give some thought to the fallen logs. Once a tree has "shuffled off this mortal coil," it certainly has more value to give. By all means remove tires and abandoned shopping carts from your neighborhood stream, but when given the choice, let sleeping logs lie.

IT'S RAINING,
IT'S POURING, THE CLIMATE
IS CHANGING

Those who have lived in New England for as long as I have might suspect that intense fall storms are becoming more common. If you feel that way, it's not your imagination. Jonathan Winter, assistant professor of geography at Dartmouth College, says that indeed, since 1996 the Northeast has received about 1.5 times the number of extreme precipitation events than in previous decades.

This figure illustrates a key challenge when it comes to our thinking about climate—how do we perceive climate change? Almost a third of us were not yet born in 1996 so it's impossible for many to remember a time before smartphones or social media, let alone environmental change on this scale. Moreover, our view of the world is often limited by where we live. For example, Hurricane Sandy caused widespread destruction in New Jersey and New York, Irene devastated Vermont, and Isaias battered Connecticut. Considered individually, each of these storms is simply an example of extreme weather. But when viewed together, these hurricanes paint a very different picture. For most people, it's as hard to observe our changing climate as it is to see the bacteria that sicken us. But unlike bacteria, scientists can't just haul out a microscope to see our changing climate.

To truly "see" what's happening to our planet, scientists rely on a variety of tools that give us a broader view. These tools include super computers and satellites, but also low-tech rain gauges employed by generations of scientists who may have never considered a changing climate. They must go back in time to understand how conditions differ today and look beyond what's happening in a particular location, such as a city or even a state, to consider regional shifts. In other words, scientists need lots and lots of data.

Dr. Winter and his colleagues indeed analyzed lots of data for their research. They crunched numbers spanning back more than a century from 116 weather stations scattered throughout New England. They looked at total rainfall and then focused their study on very large storms, the kind of storms that frequently have proper names, so-called extreme weather. But how does one define what is extreme?

Many scientists consider events that happen one time out of twenty to be exceptional; think of the 0.05 p value used by statisticians. But Dr. Winter raised the bar. He defined "extreme events" as those that happen just one time out of a hundred. In other words, he focused on the largest 1 percent of rain events. To put this in human terms, a newborn would have to weigh 10 pounds, 6 ounces to be in the top 1 percent for birth weight.

So how has precipitation from the top 1 percent of New England's storms changed since 1901? It turns out that the volume of precipitation falling on the wettest days was consistent for almost a century. But starting in 1996, precipitation began to increase. Again, taken alone, 1996—or 2024 for that matter—might just be an unusually rainy year. This is what scientists call "weather." But year after year, the volume of extreme precipitation falling in the Northeast in autumn has increased by 50 percent. This confirms that something significant is happening to our "climate."

This uptick in extreme precipitation has serious consequences for human communities. Civil engineers use one-hundred-year floodplain maps to determine where it's safe to build homes, but these maps no longer reflect reality in the era of climate change. What is the meaning of a so-called hundred-year flood if such a flood happens, say, every thirty years, or more frequently? Some parts of New England are already experiencing "hundred-year storms" every few years.

As an aquatic ecologist, I am interested in the effects of these extreme storms on water bodies. Cyanobacteria, also called "blue-green algae," thrive in warmer temperatures. Like plants, cyanobacteria need nutrients, including phosphorus and nitrogen, and they need light. These factors help us predict where and when potentially harmful algal blooms may occur. Increased rainfall from storms flushes nutrients into lakes, ponds, and other water bodies where they remain in circulation for long periods and can amplify cyanobacteria growth. And storms are not the only changes influencing water bodies. As our climate warms, data compiled by the Environmental Protection Agency show New England states are warming faster than any other region in the Lower 48.

This essay was greatly improved by fact-checking and editing by Dr. Jonathan Winter.

OUT, OUT, DAMN DAM

On most days, a significant portion of Winooski River's water flows through three six-foot turbines of the Winooski One Dam, sustainably meeting the daily electrical needs of many Burlington Vermont households. While the Winooski One Dam delivers clean power, other dams meet diverse needs including recreation, navigation, irrigation, flood control, water supply, as well as hydropower.

And while dams are serving our needs, they produce side effects, not least of which is the interruption of fish passage. Some dams include fish ladders providing migration routes. The Winooski One Dam includes a fish lift to intercept fish as they swim upstream.

Safely intercepted salmon are trucked 12.5 miles upriver to Williston. This journey gets them past two additional dams to suitable spawning habitat. In 2018, forty-two salmon made the trip, and the number has exceeded one hundred salmon in five of the past ten years.

Blocking fish migration is just one problem that dams create. Sunbaked reservoir water flowing over dams is abnormally warm and this influence can extend for miles downstream. Boris Kondratieff and Reese

Voshell of Virginia Polytechnic Institute studied the mayfly *Heterocloeon curiosum* in Virginia's North Anna River. The number of nymphs below the dam was half that of a nearby free-flowing river. In the same study the authors noticed that the second generation of the yearly mayflies hatched a full month late because the river water remained warm later into the fall below the dam.

Sediment also accumulates behind dams. Rapidly flowing river water carries much sediment that promptly settles out in relatively still reservoir waters. In May 2019, I stopped in Hartford Connecticut to stretch my legs during a road trip. I wandered the Riverside Park paths by the Connecticut River. The park was being cleaned following recent flooding, and mud washed from behind one of more than twenty Connecticut River dams exceeded two sticky feet of depth in places.

John Harrison and colleagues from Washington State University study reservoir sediments and climate change. As sediment builds and material decomposes, bacteria produce methane. When we draw reservoir water levels down, the drop in water pressure releases much of this methane, a potent greenhouse gas, to the atmosphere.

Removing dams certainly solves some of these problems. Free-flowing rivers process, transport, and deposit sediments. Much of this sediment is deposited on the river's floodplain, where it supports riparian vegetation that is essential wildlife habitat. Following dam removal, the sediment that accumulated below water may be removed, or may simply be allowed to be redistributed by the restored river flow.

According to Peter Zaidel of UMass Amherst, historically we have removed dams for three common reasons: safety, to reduce maintenance cost, and to restore fish populations. The catastrophic loss of life caused by sudden dam failure is a rare event but should certainly give owners of defunct dams pause.

Beginning in the 1980s, US dams were more frequently removed for conservation reasons. Angela Bednare from University of Pennsylvania found that dam removal improved habitat for fish, birds, and mammals as well as improving sediment transport through the river basins. On the downside, one study she reviewed documented PCB release on the Hudson River, and another mentioned loss of reservoir fish species. This begs the question: Should we expect reservoir species in a river?

According to Francis Magilligan of Dartmouth College, 1,100 dams have been removed in recent decades in the United States; 127 of those in New England. In August 2009, a dam was removed from the Sedgeunkedunk Stream that flows into the Penobscot River in Maine. Robert Hogg from the University of Maine and his colleagues noticed that fish diversity and abundance quickly increased above the former damsite, and Atlantic salmon spawned there for the first time since dams were installed in the 1800s.

Similar success stories abound, and yet it would be foolhardy to suggest removing all dams everywhere. Many, including the Winooski One Dam, are early in their useful life and provide cleaner energy than other alternatives. With more than 80,000 dams on the US Army Corps of Engineers National Inventory of Dams, and tens of thousands of smaller, non-inventoried dams, we'd certainly have our work cut out for us.

Thousands of dams were not built in a day and won't be removed all at once. But many dams can and should be removed. We can't solve all fish migration problems by truck, and a great many dams no longer serve their original purposes. With careful cost-benefit analysis and planning, we can make a dent and return more rivers to their natural, free-flowing states.

COBBLESTONE TIGER BEETLES

Earlier this summer, I joined graduate school friend and beetle biologist Kristian Omland in search of the elusive cobblestone tiger beetle (*Cicindela marginipennis*). We loaded a canoe with insect nets, jars for photography, and binoculars to view beetles while minimizing handling. Absent from our kit: entomologist's killing jars. Ours was a catch-and-release mission. The cobblestone tiger beetle is a species of greatest conservation need (SGCN), and we certainly would do nothing to lower its numbers.

The cobblestone tiger beetle is a half inch long, brown to olive green, lanky, fast-moving insect. The elytra covering its flying wings are bordered by a cream-colored scalloped stripe. When it raises its elytra to fly, this beetle reveals a fire-engine-red abdomen. Long legs keep the beetle off sunbaked stones, and dense white hairs on its underside reduce radiant heating from below.

Adult tiger beetles run and fly rapidly to chase down and subdue smaller insects. They use long, sickle-shaped, toothy mandibles to catch and perforate prey insects, then release enzymes and acids strong enough to put holes in an entomologist's net. This results in a soupy meal, which the beetle eats before discarding its prey's empty husk.

Larvae use a different hunting strategy. They live in vertical tunnels in soil and use their armored and camouflaged heads and first thoracic segments to plug the entrances to these tunnels. Should an ant or other small insect get close enough, the larva grabs the prey using pincer-like mandibles and drags it below ground to make a meal of it. The larvae use a hooklike abdominal segment at the back end to anchor themselves within the tunnel, reducing the chances that large prey will yank them from their burrow.

Habitat loss is the most likely factor leading to this species' inclusion on the regional SGCN list. Each tiger beetle species has a particular niche, and as their name suggests, cobblestone tiger beetles favor cobble-strewn beaches, islands, and gravel bars in large rivers. Females use long ovipositors to make holes in the soil and lay their eggs. Once hatched, larvae enlarge holes by tossing out sand and silt; the resulting "throw piles" of debris can be helpful in locating larval burrows.

Protecting cobblestone beaches along the Connecticut, the Winooski, and other rivers where we know these beetles live is likely the best strategy to preserve these beautiful insects. And river dams are the single largest threat to these habitats. Free-flowing rivers move across their floodplains building, eroding, and rebuilding sand and gravel bars, beaches, and islands such that cobblestone features in rivers vary in age from freshly formed to long-established. Beaches of different ages provide different microhabitats for a range of organisms and increase biological diversity in river corridors.

New beaches lack vegetation. Gradually, plants such as dogbane move in and stabilize substrates. The next successional stage includes willow saplings and cottonwoods that in time grow to be substantial trees. Occasional Tropical Storm Irene–scale floods reset the system by removing vegetation and redepositing sediments. Receding floods first drop the largest rocky sediments and cobbles, later the gravel, followed by sand, and eventually silt. This sediment sorting leaves cobblestone beaches high and dry above the waterline—creating prime real estate for cobblestone tiger beetles.

Dams hinder this natural beach rebuilding process, thereby eliminating several microhabitat types for significant distances upstream. This eradicates habitat for cobblestone beetles and for a whole set of other species, including plants, insects, and the birds and fish that rely on them for shelter and food.

Kristian and I searched cobblestone beaches by walking near dogbane patches and on newer plant-free beaches and mid-channel bars. The first gravel bar we visited was at the confluence of the Huntington and Winooski rivers, where Jonathan Leonard, coauthor of *Northeastern Tiger Beetles*, and his daughter Emma found cobblestone beetles in 1997, and where Kristian has seen them since.

We were rewarded with tiger beetle sightings, but of a different species. The common shore tiger beetle (*Cicindela repanda*) is less picky about habitats and was the only tiger beetle species we saw on our trip. Other scientists have observed cobblestone tiger beetles later in the summer, and so I think additional canoe trips are warranted. I'm happy to have any excuse for more time on the river.

TIME TRAVEL IN A PEAT BOG

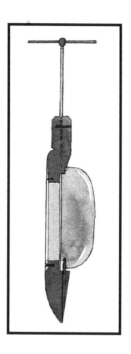

G utter pipes full of soggy peat show up on the bench by my office each March. This means one thing: my colleague Peter Hope's Saint Michael's College students are about to experience time travel. You might reasonably ask how pipes filled with peat could possibly relate to time travel. What? No DeLorean, flux capacitor, or 1.21 gigawatts of electricity? To answer, we need to consider where peat comes from and how it forms.

Peat accumulates in bogs over millennia. Decomposing plant material consumes oxygen, and sphagnum moss turns water acidic by pulling minerals from the water and releasing acid. When dead plants and moss pile up in acidic water with little oxygen, they remain more or less preserved. The resulting accumulation is called "peat."

Pollen accumulates along with the peat, and that is where the time travel comes in. Pollen falls into two broad categories: the familiar sticky stuff carried by insects and wind-blown pollen that makes us sneeze. Trees cast vast quantities of pollen on the wind and a few grains hit their intended targets. Far more pollen blows around the landscape, and some of it ends up in bogs.

Two pollen grain traits make Hope's time travel experiment possible: they have distinctive appearances that identify the type of plant each grain came from; and they are so resistant to decay that they last as long as the bogs in which they accumulate. At the top of the bog, we find pollen from trees still growing today; as we go deeper, we travel back in time and learn what trees surrounded the bog in times past.

When less hardy people are comfortably indoors, Hope and his students drive long metal pipes into a Vermont bog. He told me that lightning is the only weather that would give him pause; handling tall metal pipes during electrical storms is ill advised.

The business end of his peat-boring device is like an apple corer. A trapdoor runs along its length. After driving the core down into the bog, a quick twist closes the trapdoor, and the contained peat can be extracted like a cork from a bottle. After students remove the first peat core and safely stash it, they drive the coring device deeper into the same hole, further back in time.

With a dozen vertical feet of peat, students have traveled 12,000 years into the past, when woolly mammoths, mastodons, and saber-tooth cats were headed to extinction; all without the use of Doc Brown's souped-up sports car. The carefully labeled peat samples record the ghosts of forests past.

It will surprise few to learn that the pollen grains found close to the surface include spruce, hemlock, and pine: all trees common at lower elevations in the Champlain Valley. Things get more interesting as the students delve deeper.

About 20 percent of the pollen from deep in the bog came from balsam fir, a rare tree in today's lowland Champlain Valley. To find a lot of balsam fir today, you'd need to go uphill, quite a way uphill. Elizabeth Thompson and Eric Sorenson's *Wetland, Woodland, Wildland: A Guide to the Natural Communities of Vermont* suggests seeking spruce-fir forests at elevations above 2,500 feet. Fir's range has moved quite a distance from the 320-foot elevation of the bog our students visit.

Fir and other tree species continue to move. Research led by the University of Vermont's Brian Beckage has chronicled tree movement over the last forty years in the Green Mountains. In his study blocks, fir trees at lower elevations have declined, while seeds that fell above the species' historic upper limits have germinated and flourished. There's evidence that other tree species are also shifting their populations up-hill. The changes in pollen our students observe send a clear message: forest composition has changed slowly but surely over time, and these changes correlate with changes in the climate. Brian Beckage's recent work paints a picture of far more rapid change.

Hope's students need spend only a few hours in the lab to get a sense of what forest composition was like 12,000 years ago. During their lifetimes they will likely see additional changes and shifts in tree distributions that have already led to the redrawing of plant hardiness zones in our region.

These are essential lessons for all of us. Climate is changing rapidly, which affects forest composition. It will be up to us to determine how we will respond and adapt to these long-term changes.

INVADING ZEBRAS

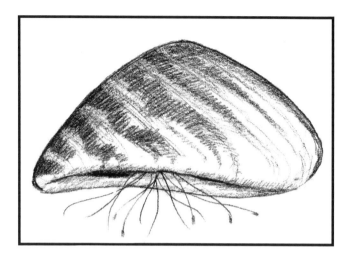

Invasive species have very well earned their bad reputations. English sparrows compete with native birds from Newfoundland to South America. Australian brown tree snakes are well on their way to exterminating every last bird from the forests of Guam. And I don't think anyone can fully predict how Colombia's rivers will change in response to drug lord Pablo Escobar's escaped hippopotamus population.

While our climate protects us from rampaging hippos, the Northeast has plenty of exotic species in its waterways, including some that cause serious damage. Zebra mussels are possibly the most familiar of these. They were first discovered in Lake Champlain in 1993 by a precocious fourteen-year-old, Matthew Toomey, who recognized one based on an identification card he'd received at school. Since then, the mussels have spread throughout the lake, and their effects have been well chronicled. They kill native mussels; coat surfaces with razor-sharp shells; foul anchor chains; block water-intake pipes; and steal plankton and other food from native mussels and fish.

With all of the negative press regarding the species, you might find it jarring to read anything positive about zebra mussels, particularly anything written by a biologist.

Discussing positive effects of invaders is practically taboo. We don't speak ill of the dead; we never praise invasive species. I'm certainly not advocating zebra mussel propagation, but like them or not, they are here to stay and perhaps the best thing we can do is limit their spread. These mussels are an important part of European ecosystems, and it's interesting to consider what native organisms benefit from their presence.

Zebra mussels are voracious filter feeders. A single mussel can suck a liter of water through its body daily. All of this filtration removes plankton and particles from lake waters, but these particles don't just disappear. The phrase immortalized in Tarō Gomi's children's book *Everyone Poops* applies. Along with excrement, unpalatable particles rejected by zebra mussels are mixed with mucus and dropped on the lake floor. Mussel excrement and mucus might not sound appetizing, but it's a smorgasbord for lake floor invertebrates and fish.

In addition to covering rocky surfaces, zebra mussels often carpet lake floor sand and silt. Formerly soft sediments that provided foraging grounds for sturgeon and other fish can become a tangled mess of living and dead mussels several inches thick. Not surprisingly, fish such as log perch, bullhead, and sculpins have difficulty finding their insect prey among the clutter of shells layered over their sandy habitats. When given the choice, the juvenile sturgeon avoid zebra mussels and spend their time on sandy or rocky areas.

What's bad for these predators may be good for their prey. To figure out just how good or bad zebra mussels could be for Lake Champlain invertebrates, we ran experiments under thirty feet of water in sandy areas of Appletree Bay. When my colleagues Ellen Marsden, Mark Beekey, and I fenced off lake floor patches with and without zebra mussels, twice as many invertebrates colonized areas with zebra mussels. More species also moved in. After a month, the number of species in experiments with added mussels doubled and included some species more typical of rocky lake floors. Nooks and crannies between zebra mussel shells seem to act like very small natural shark cages that protect tiny insects from hungry fish. And when we placed insects and hungry insectivorous fish in aquariums, far more invertebrates survived with zebra mussels than without.

Since running these initial laboratory experiments, my students have confirmed the patterns time and again using samples taken from several locations in Lake Champlain. Dredge samples taken from Missisquoi Bay, Saint Albans Bay, and Lake Champlain's Inland Sea often contain zebra mussels. And samples with more zebra mussels invariably contain more invertebrates along with the striped mollusks.

On balance, I would rather have a lake without zebra mussels than with them. But unless ways are found to eliminate them, it will remain important to understand how they affect native species. In Lake Champlain, the zebra mussel population grew rapidly and has since fallen below peak numbers, as often happens with this species in a new location. In 2015, for the first time in some years, we pulled up a lake floor sample from the shallows of Burlington Bay that entirely lacked zebra mussels. Perhaps we are reaching a new equilibrium?

GLOSSARY

I have tried to avoid jargon in this book, but some terms are essential to the understanding of aquatic ecology, and this brief glossary explains many of these terms.

Aphotic zone—the portion of a lake's "water column" that lacks light. The aphotic zone's upper boundary is defined as the depth to which less than 1 percent of the light reaching the lake's surface penetrates and extends from there to the lake floor. Ponds and shallow lakes may lack this zone; see "photic zone" and "Lakes: Life in Standing Water."

Basin—land area from which water drains to a single stream or river. Smaller basins feeding river tributaries are called "subbasins" nested within larger river basins; see "watershed."

Beaver meadow—a grassy/wildflower community that grows in the rich soil that accumulates in former beaver ponds.

Benthic zone—the floors of any and all aquatic habitats.

Benthos—the community of organisms inhabiting the benthic zone.

Bioturbation—organisms digging, tunneling, and disturbing accumulated soils and sediments. When this happens underwater, the activity can move particles from the "benthos" into the "water column."

Boundary layer—a very thin layer of water in contact with and immediately adjacent to river and stream rocks where friction has reduced water velocity to near zero; importantly, turbulence may still exist absent net water movement; see "Flat as a Pancake."

Catchment—see "basin."

Channelized—when streams and rivers are straightened and constrained into narrow channels to accelerate water downstream; contrast with "incised"; see "daylighting."

Coarse woody debris—tree trunks and branches that provide important habitat and influence sediment deposition and river meander formation. In very large rivers this material moves downstream with the current; see "Let Sleeping Logs Lie."

Compensation point—the depth in a lake at which the aquatic community's combined metabolic activity exactly consumes all the oxygen being produced by photosynthesis; see "aphotic zone" and "photic zone."

Complete metamorphosis—insect life cycles that include egg, larva, pupa, and adult forms; note that the term "larva" is used rather than "nymph"; see "incomplete metamorphosis."

Convergent evolution—when two or more unrelated organisms develop similar features; see "If It Looks Like a Snail, It Might Be a Caddisfly."

Countershading—pattern of darker colors on organisms' upper surfaces and paler colors beneath that improves camouflage against dark backgrounds when viewed from above; and against the sky when viewed from below; see "Summer Skaters" and "Upside-Down Aquatics."

Cultural eutrophication—when nutrients, and particularly phosphorus, originating from human activities enter water bodies, leading to "eutrophication."

Daylighting—removing streams from underground pipes so that they can once more flow on the land surface.

Density anomaly of water—unlike most substances, water is most dense at 39.2 degrees Fahrenheit (4 degrees Celsius); when water is cooled past that point it expands whereas most other substances continue to shrink on cooling; see "Life at 39 Degrees" and "Ice Capades."

Detritus—organic material that accumulates on the landscape and in water bodies; see "The Afterlife of Leaves."

Ecosystem engineer—an organism that physically modifies its habitat impacting other species in its community; see "Landscape Engineers" and "The Smallest Engineers."

Ecosystem service—any benefit provided by the natural environment; monetary values can be calculated for such services to illustrate replacement cost absent the intact environment. Examples include flood control, cleaning the water supply, and pollination; see "Water Cleaned for Free" and "Wetlands Reduce Flooding and Improve Water Quality."

Emergent lifestyle—insect life cycles that include larval/nymphal stages underwater and adults that disperse and reproduce out of the water; see essays in part 5, What Emerges from the Deep, and see "non-emergent lifestyle."

Epilimnion—the warmer, upper layer of water in a stratified lake; see "hypolimnion" and "thermocline."

EPSCoR—Established Program to Stimulate Competitive Research, a National Science Foundation program to improve research competitiveness in targeted US states and territories. Ideas referenced in this book stem from Vermont EPSCoR research, and all of the essays contribute to Vermont EPSCoR's high school outreach program; see "Ice-Out and Climate Change"; "It's Raining, It's Pouring, the Climate Is Changing"; and "Let Sleeping Logs Lie."

Euphotic zone—see "photic zone."

Eutrophication—excess plant and algal growth in lakes and ponds that reduces oxygen concentration, particularly when photosynthesis stops at night and plant/algal/animal respiration consumes oxygen leading to fish kills. Eutrophication occurs naturally as ponds and lakes age; but see "cultural eutrophication."

Facilitation—when the presence of one species in a habitat helps or increases the abundance of one or more other species. Often, this is because of a physical change in the environment; see "ecosystem engineer," "Landscape Engineers" and "The Smallest Engineers."

Filtering collectors—macroinvertebrates that feed by extracting fine particles of organic matter from the water column; see "Functional feeding groups," "Life, Death, and Blackflies"; "Rivers: Life in Flowing Water"; "The Afterlife of Leaves"; "Submerged Silk Spinners"; "Burrowing Mayflies"; and "Invading Zebras."

Frass—macroinvertebrate excrement, a food resource for "filtering collectors."

Functional feeding groups—macroinvertebrate categories morphologically and behaviorally adapted to feed on specific food resources; see "filtering collectors," "gathering collectors," "predators," "scrapers," "shredders."

Gathering collectors—macroinvertebrates that feed by gathering fine organic material that has settled to the bottom of water bodies; contrast with "filtering collectors"; see "functional feeding groups."

Glochidia (singular glochidium)—microscopic larval life stage of many mussels and clams that attach to fish.

Grazers—see "scrapers."

Groundwater—water found below the ground surface in the minute spaces between soil particles and in cracks and pores in rock; see "water table."

Gulf Stream—an oceanic current originating in the area of the Gulf of Mexico and flowing northeast along the east coast of the North American continent; see "An Extraordinary Fondness for Heat."

Hardness—mineral content of water, particularly of calcium and magnesium; reduces lather formation and causes mineral deposition in mechanical devices such as dishwashers; see "water softener."

Hypolimnion—the colder, lower layer of water in a stratified lake; see "epilimnion" and "thermocline."

Hyporheic zone—groundwater extending below and from the sides of rivers and streams. Water exchanges between the hyporheic and the river, so oxygen content may be higher than in groundwater some distance from a river. Some macroinvertebrates spend part or all of their life cycle in the hyporheic; others use this zone as a storm refuge; see "Keeping It Clean Downstream."

Impervious surfaces—roofs, driveways, road surfaces, and parking lots that prevent rainwater infiltration to groundwater, instead directing it to streams causing bank and bed erosion; see "pervious surfaces" and "Keeping It Clean Downstream."

Incised—when the water flow in small streams is increased, frequently due to "impervious surfaces" causing vertical erosion and sinking the stream into the landscape.

Incomplete metamorphosis—insect life cycle including egg, nymph, and adult stages but lacking the pupal stage; note that the term "nymph" is used rather than "larva"; see "complete metamorphosis."

Larva(e)—the second life stage of many insects. Larvae tend to be quite different than adult insects; dipteran (true fly) larvae for example entirely lack legs; contrast with "nymph"; see "complete metamorphosis."

Lentic—standing water habitats such as ponds, lakes, vernal pools, and water-filled tree holes.

Limnetic zone—open-water area of larger, deeper lakes lacking rooted vegetation. This zone extends from the surface to the bottom of the photic zone and may be absent from shallow lakes and ponds; see "profundal zone."

Littoral zone—shallower portion of lakes where vegetation is rooted in the benthos.

Lotic—flowing water habitats such as rivers, streams, springs, and seeps; contrast with "lentic"; see "Rivers: Life in Flowing Water."

Macroinvertebrate—invertebrate organisms large enough to be seen with the naked eye. Scientists use mesh sieves to extract these larger organisms from samples; 1 mm sieves were commonly used, but many recent studies use 0.25 mm to retain midges and mites. Studies in this book adopted Vermont DEC's 0.6 mm standard.

Match the hatch—anglers attempting to ensure that artificial flies best represent insects currently emerging from the body of water they are fishing in; see "prey switching" and "Burrowing Mayflies."

Metalimnion—see "thermocline."

Neuston—the community of organisms living on, and supported by, the surface tension of water bodies, including those organisms hanging beneath the surface tension; see essays in part 2, Life on Top.

Non-emergent lifestyle—life cycles of invertebrates spent entirely underwater and not emerging as adults; contrast with "emergent lifestyle."

North Atlantic Current—an oceanic current fed by the Gulf Stream and flowing northeast from the east coast of the North American continent to northwestern Europe; see "An Extraordinary Fondness for Heat."

Nymph—the second life stage of many insects. Nymphs tend to be morphologically relatively similar to adult insects with the major difference being the lack of wings; contrast with "larva"; see "incomplete metamorphosis."

Oxbow—a lake or pond formed in a former river loop or meander after the river channel has moved to a different location. Older oxbows can become riparian wetlands as they infill; see "Wetlands Reduce Flooding and Improve Water Quality."

Periphyton—the complex community of algae, fungi, bacteria, and other organisms coating submerged surfaces such as rocky riverbeds; see "scrapers."

Pervious surfaces—driveways, sidewalks, and other components of the built environment that permit water percolation to groundwater; contrast with "impervious surfaces" and see "Keeping It Clean Downstream."

Photic zone—the portion of a lake's water column that light penetrates. The photic zone's lower boundary is defined as the depth at which less than 1 percent of the light reaching the lake's surface penetrates; see "aphotic zone."

Pleuston—often used as a synonym of "neuston" but also used to refer to the microscopic portion of the neuston, or the very small organisms living on the surface film.

Pool—an area of a stream where the water surface is relatively flat and where there tends to be net deposition of material. Step-pools separated by steep cascades form in high-gradient mountain streams; pools occur in riffle-pool sequences in medium gradient streams.

Predator dilution—the concept that any one animal in a large group is less likely to suffer predation because predators have many other prey from which to choose; see "Four Eyes on You."

Predators—the macroinvertebrate "functional feeding group" that specializes in eating other invertebrates; see "Underwater Assassins."

Prey switching—the tendency among predators to specialize on the most abundant prey until a different prey species becomes more common; see "match the hatch" and "Burrowing Mayflies."

Profundal zone—the deep portion of a lake lacking light. It extends from the bottom of the limnetic zone to the lake floor; see "limnetic zone."

Pro-glacial lake—a meltwater-fed freshwater lake that forms between the receding edge of a glacier and its terminal moraine (accumulated rock, soil, and debris pushed ahead of the glacier and deposited where the glacier reached its farthest extent).

Riffle—an area of steeper, more turbulently flowing water that usually occurs as part of riffle-pool sequences found in streams of medium gradient. Riffles have net erosion of materials and therefore have cleaner, more open spaces between rocky substrates.

Riparian zone—the area of vegetation adjacent to a river or stream. The area is influenced by the stream characterized by shrubs, herbs, and trees that can tolerate occasional inundation; see "Wetlands Reduce Flooding and Improve Water Quality."

Rocks—the term "rocks" is used throughout the book as a convenience. However, in scientific parlance, "rock" lacks precision, and ecologists use more precise categories borrowed from geologists and engineers. Roughly speaking, "gravel" refers to substrates in the size range of a marble; "cobble" applies to fist-sized rocks; rocks approaching the size of your head or mine are called "boulders," and things larger than that that don't move much in stream flow are called "bedrock."

Scrapers—the functional feeding group that makes a living by grazing periphyton from rocky surfaces.

Secondary compounds—plant-produced chemicals that are not directly essential to the plant's metabolism, survival, growth, and reproduction. These chemicals including various toxins that discourage herbivory and remain in leaves that fall into aquatic systems.

Shredders—the macroinvertebrate functional feeding group that consumes fallen leaves in aquatic habitats.

Subbasin—explained under "basin."

Thalweg—the deepest part of a river's flow between the river's banks. The thalweg may be at the center of the stream or closer to one bank as the river twists, turns, and erosion varies across its width.

Thermocline—a zone of rapid temperature change between the warmer, upper layer, and the colder, lower layer of a stratified lake; see "epilimnion" and "hypolimnion."

Vernal pool—a pond that dries out at some point most years. Because bullfrog tadpoles and fish require water year-round, they do not occur in vernal pools, allowing for different macroinvertebrate and amphibian communities to persist absent these predators.

Water column—the water between the surface and the floor of a water body. Contrast with "benthic zone."

Watershed—this word is synonymous with "catchment" and "basin," particularly as used in the United States. However, it can also mean the boundary between two catchments, often on a mountain ridge that sheds water on one side to one catchment and to a different catchment on the other side.

Water softener—a device that removes the calcium and magnesium ions from water to reduce "hardness." The removed ions are usually replaced with sodium.

Water table—the upper level of "groundwater" at a given location.